TALKING

A VOLUME IN THE

COLLECTION ON TECHNOLOGY AND WORK

edited by Stephen R. Barley

Talking about Machines

An Ethnography of a Modern Job

JULIAN E. ORR

ILR PRESS *an imprint of*

CORNELL UNIVERSITY PRESS   *Ithaca and London*

First published 1996 by ILR Press/Cornell University Press.

Printed in the United States of America

♾ The paper in this book meets the minimum requirements
of the American National Standard for Information Sciences—
Permanence of Paper for Printed Library Materials, ANSI Z39.48-1984.

Library of Congress Cataloging-in-Publication Data

Orr, Julian E. (Julian Edgerton), 1945–
    Talking about machines : an ethnography of a modern job / Julian E. Orr.
        p.   cm.—(Collection on technology and work)
    Includes bibliographical references and index.
    ISBN 0-8014-3297-9 (cloth : alk. paper)—ISBN 0-8014-8390-5 (pbk. : alk. paper)
    1. Xerox Corporation—Customer services.   2. Photocopying machines—United States—Maintenance and repair.   3. Mechanics—United States.   4. Ethnology—United States.   I. Title.   II. Series.
HD9802.3.U64X4765   1996
305.9′6864—dc20   96-23924

*To the men and women
working in field service*

Contents

Foreword

BY STEPHEN R. BARLEY

Pick up a newspaper and turn to the business section. Skim a recent newsmagazine. Browse the nonfiction best-seller rack at your local bookstore. Chances are you will easily find one or more discussions of how the economy as we have known it is being swept away by waves of change. Over the last two decades, commentators have penned tens of thousands of books and articles on the transformation of industrial society. By now the contours of the change are well known, although the scope of the ramifications remains unclear. Competition is becoming increasingly global and corporations increasingly multinational. Computers, advanced telecommunications, and other digital devices are reconfiguring the technical infrastructure. Mergers and acquisitions have led to the consolidation of almost every industry. The blue-collar labor force is rapidly disappearing, and even clerical work has begun to wane. Employment is shifting to the provision of services, broadly construed. Firms are dramatically reducing the size of their workforces. Even the once sacred jobs of midlevel managers are no longer immune to being "re-engineered." Since the turn of the century, professional and technical workers have moved from the most peripheral to the largest occupational sector in most industrialized nations. Contingent and part-time work is growing by leaps and bounds, even among the professional and managerial labor force. The hope of lifelong employment is quickly becoming a relic of the past.

There is no lack of speculation about what all of these changes mean, especially for organizations. Many a fortune has been made advising firms how to change their strategies and structures to survive the brave new economy. Nor is there a dearth of advice for individuals, especially those in the managerial ranks. An entire genre of self-help writing has arisen in the business press that advises people to seek lifelong learning, to take charge of their own careers, and to become, as one recent book put it, entrepreneurs of themselves. But despite all the speculation, discourse, and guidance, one conversation is notably but a murmur: what do people do in this new economy? Amid the dust of the rush to downsize, re-engineer, compete, compute, empower, and predict, work has almost disappeared from sight. There is more than a little irony here.

History tells us that work is the bedrock of any socioeconomic system. When a society's mode of production changes, so does the nature of work. It is primarily for this reason that the industrial revolution warrants being called a revolution. The industrial revolution marked a shift in what people did for a living and how they accomplished tasks. It signaled the decline of agriculture and handicraft and the rise of factory and office work as the primary means of making a living. Out of the crucible of the second industrial revolution (the late 1800s) came the time clock, the corporation, the union, the occupation of management, and even the very idea of "having a job" or of stringing those jobs together into a career. There is absolutely no reason to believe that changes in the nature of work have become any less crucial to the dynamics of a socioeconomic shift than they were in the nineteenth century.

For instance, how can a firm effectively reorganize or re-engineer its operations without understanding the work that its employees do? The obvious answer is, it can't. Yet lack of knowledge does not appear to stop organizations from trying. How can computers and microelectronics change the economy or restructure the way organizations do business without changing the nature of work? The obvious answer is, they can't. Yet this fact is hard to detect in the burgeoning literature on information systems. From the pages of the *MIS Quarterly* to *PC Magazine*, the computer revolution is typically fought in a black box where we never learn what people do, only that they should now be able to do whatever they do faster and more easily by computing. What meaning can the "service economy," the "information economy," the "knowledge economy," and similar terms have unless they denote substantive changes either in what people do for a living or how they do it? The obvious answer is, very little. Yet journalists, futurists, and even sociologists routinely employ such epithets without explaining precisely what kinds of work they have in mind. In fact, if one looks carefully at how these terms are used, one discovers that they seem to cover just about any kind of work except blue-collar work and farming. They seem to be little more than trendy synonyms for "white-collar." The upshot is that millions of people go to work each day to do things that almost no one but themselves understands but which large numbers of people believe they know enough about to set policy, offer advice, or redesign. Work has become invisible.

In the opening chapter of this book, Julian E. Orr suggests that work's invisibility reflects the fact that it has become an abstraction, a generalized input into a production function. This was not always the case, and there are a host of reasons for the change. In the past, when occupations and work were less differentiated, it was easier for people to know what other people did because there was simply less to know. In the past, those who ran organizations were familiar with the production processes. They often designed the process and had even done the work themselves. Today, organizations are so complex that it is difficult for those in charge to have experienced much of the organization's work firsthand. Moreover, managers are often hired from the outside, and their experience frequently lies in completely different industries. Even industrial engineers know less about work than they once did. Whatever else one might think about Frederick Taylor and the other founders of human factors research, they cannot be faulted for failing to observe work processes. They did not reconfigure factories or assess efficiency on the basis of abstract theory, financial indicators, or computer simulations; instead, they attempted to look at the specifics of work in considerable detail.

Scholarly developments have also conspired to make work more invisible. Prior to the 1950s, organizational scholars relied heavily on case studies and field research. Industrial sociologists and students of what used to be called social organization viewed the study of work, occupations, and organizations as intimately related. When organizational theory broke from industrial sociology in the 1960s, its founders abandoned the study of work to solidify their jurisdictional claim. Work became the intellectual property of its sister discipline, the sociology of work and occupations. Because early organizational theorists were committed to seeking general principles of organizing, when it was necessary to talk about work, they turned to abstractions. With concepts like complexity and uncertainty, researchers hoped to level distinctions between work as dissimilar as management and medicine in order to discover relations that would hold across contexts. Over time, most sociologists of work and occupations also turned to large-scale quantitative studies that focused on such issues as social stratification and occupational prestige. Thus, field research on work ironically began to decline on all fronts at precisely the time that the occupational structure began to shift.

The combination of social and academic trends has led us to a situation where we know more about yesterday's work than we do about today's. The predicament is troubling because one can develop neither policies nor theories without at least implicit models. Policymakers, managers, consultants, union organizers, and academics therefore frequently fall back on images of work based on the occupational structure and industrial culture of the first half of the twentieth century.

The evidence is overwhelming. Managerial theorists and the business press routinely write as if all managers do the same sort of work, even though management has become highly differentiated. The union organizer generally assumes that technicians will respond to the same bread-and-butter issues that were so successful in organizing the blue-collar labor force, yet they do not. Leftists routinely orient themselves to service work by assuming that service workers are a new proletariat even though most service workers apparently don't see themselves this way. Occupational sociologists and MIS specialists write as if clerical and secretarial work has changed little since the 1950s, except that computers have been substituted for typewriters and adding machines. Even the most casual observation of what people do and how they think about their work calls all such claims into question. Unless we begin to examine what people in modern jobs actually do, we run the risk of generating theories and policies that not only lack verisimilitude but may actually prove to be pernicious. It seems unreasonable to believe that people can plan, manage, organize, or even write intelligently about what they don't understand.

One way out is once again to examine work in context and to reward those who do. We sorely need rich descriptive data on what people do and how they do it, not only because such data will improve our theories and our decisions but because only with such information can we develop an appreciation of and respect for the contributions that people make each day to the society and economy in which we live. It is with this goal in mind that the ILR Press has established the Collection on Technology and Work, of which this is the first volume. The collection will be a home for research, especially ethnographic research, that helps us better understand how the nature of work is changing. Our aim is to publish books that help make what people do all day once again visible.

I can think of no better book with which to launch the collection than this one. Julian Orr's study of photocopier repair technicians at Xerox has for some years now been an underground classic among ethnographers of work. First completed as a dissertation in 1990, that version was distributed as a "Blue and White," a Xerox PARC technical report. Even with this limited distribution, the book you are about to read has been influential; it is the source of a number of ideas that have recently gained considerable currency. For instance, it is here that Orr documents and develops the important and counterintuitive notion that technical knowledge is best viewed as a socially distributed resource that is diffused and stored primarily through an oral culture. Viewed from this perspective, the technicians' war stories become texts, not only for the ethnographer, as the postmodernists would have it, but for the technicians themselves. It is here, too, that Orr puts the flesh of everyday life on Lave and Wenger's idea of a community of practice,[1] an idea that promises to contribute significantly to both occupational and organization studies because it

---

[1] J. Lave and E. Wenger, *Situated Learning: Legitimate Peripheral Participation* (Cambridge: Cambridge University Press, 1991).

enables us to talk about occupational dynamics in situations that lack the institutional supports that sociologists normally attribute to recognizable occupations. *Talking about Machines* also has the distinction of being the first book-length ethnography of technicians' work ever written.

Talking about Machines demonstrates beyond a shadow of a doubt the benefits of focusing an uncompromising ethnographic eye on work practices. It should serve as a model for the kinds of ethnographies we need if we are to acquire a grounded appreciation of what work in a postindustrial economy is like. We learn from this book that technicians' work is not what their managers believe it to be. I suspect most readers will also find that photocopier repair is very different from their preconceptions, and I am sure that the next time you encounter someone repairing your photocopier, you will see both the work and the worker differently. This is because Orr shows us the dignity, the intelligence, the skill, and the dedication that photocopier technicians bring to their work. He rescues what they do and who they are from invisibility by showing us a piece of their world; by portraying their world, he shows us an image of who we are and where we are going.

Acknowledgments

Writing a book creates many debts, some of which must be acknowledged. I want to thank Carol Greenhouse, Davydd Greenwood, and Stephen Barley for their support and encouragement in the early stages of this project. Lucy Suchman patiently read the many drafts and always had pertinent comments and interesting questions. John Seely Brown gave me the opportunity and the support necessary to do the research.

Two people made enormous contributions to the progress from manuscript to book. The first is Frances Benson, director of ILR Press and now editor in chief at Cornell University Press; her unalloyed enthusiasm for the book revived mine on many occasions. The second is Stephen Barley, reappearing as series editor of this volume and still reading it after many iterations. His detailed comments and criticisms have greatly improved the book, and he can hardly be blamed for the things I would not do.

Finally, Willie Sue, my partner, wife, love, has been endlessly supportive of this endeavor, and I am forever grateful.

A portion of Chapter 8 was previously published as "Sharing Knowledge, Celebrating Identity: War Stories and Community Memory in a Service Culture," in *Collective Remembering: Memory in Society*, edited by David S. Middleton and Derek Edwards (Newbury Park, Calif.: Sage, 1990).

This work was partially funded during fieldwork by the Army

Research Institute under contract number MDA903-83-C-0189. All other support was provided by the Palo Alto Research Center of Xerox Corporation.

J. E. O.

Pescadero, California

TALKING ABOUT MACHINES

1

Introduction

Work is a constant part of our lives in the United States and other modern industrialized countries; we spend a significant portion of our lives doing something, usually for someone else, in order to earn our living. Life at work is a staple in our conversation, but we rarely talk about what we really do in the doing of the job. This omission extends to the professional literature on work: most such literature is not concerned with work as practice, by which I mean that these writings do not focus on what is actually done in accomplishing a given job. Instead, most are centered on work as the relation of employment or on work as a source of the worker's identity. Although such writings are inevitably based on assumptions about practice, practice itself is usually taken for granted, and the basis of the assumptions remains implicit. In contrast, I argue that a study of practice itself shows work to be generally different from and frequently more complex than is usually assumed; thus, a careful examination of work practice will deepen our understanding of both the relations of employment and the role of the work in the constitution of the worker's identity.

In particular, this study examines the practice of experienced technicians maintaining photocopiers for a major U.S. corporation and finds their practice to be a continuous, highly skilled improvisation within a triangular relationship of technician, customer, and machine. Technical service work is commonly conceived to be the fixing by rote procedure of uniform machines, and routine repair is indeed common. However, individual machines are quite idiosyncratic, new failure modes appear continuously, and rote procedure cannot address unknown problems. Technicians' practice is therefore a response to the fragility of available understandings of the problematic situations of service and to the fragility of control over their definition and resolution. Understanding is fragile in that accurate information about the state of the machine is only sometimes available, and the meaning of available information cannot always be found. Control is fragile both because the technicians come to work when the relationship between customer and machine is already askew and because the technicians cannot keep the machines working and the customers satisfied; they can only restore that state after the fall. Work in such circumstances is resistant to rationalization, since the expertise vital to such contingent and extemporaneous practice cannot be easily codified.

Narrative forms a primary element of this practice. The actual process of diagnosis involves the creation of a coherent account of the troubled state of the machine from available pieces of unintegrated information, and in this respect, diagnosis happens through a narrative process. A coherent diagnostic narrative constitutes a technician's mastery of the problematic situation. Narrative preserves such diagnoses as they are told to colleagues; the accounts constructed in diagnosis become the basis for technicians' discourse about their experience and thereby the means for the social distribution of experiential knowledge through community interaction. The circulation of stories among the community of technicians is the principal means by which the technicians stay informed of the developing subtleties of machine behavior in the field. The telling of these narratives demonstrates and shares the technicians' mastery and so both celebrates and creates the technicians' identities as masters of the black arts of dealing with machines and of the only somewhat less difficult arts of dealing with customers. Talk about machines is perhaps to be expected in such a job, but recognition of the instrumental nature of such talk provides a new perspective on the work.

The technicians distinguish stories told in the course of finding a solution to a machine problem from those told for purposes of boasting or idle amusement, even though the stories may be indistinguishable in and of themselves. Narratives in the latter category are characterized as "war stories," a term that connotes something useless and boring; it suggests a tolerated relative, continually telling the same stories of a war long past. Although some of the war stories are clearly intended more to amuse than enlighten, many others differ from those told in the situation of doing the work only in the context of their telling. Consequently, I do not separate war stories from other stories as the technicians do, preferring instead to distinguish the contexts in which different stories are told.

The work of technical service involves the community of technicians, the community of users, and their respective corporate entities in addition to the machines, and it occurs in a public arena, the customer's place of business. The work is analyzed here as a triangular relationship among the technicians, customers, and machines. This analysis is based Bruno Latour's assertion (1986, 1988) that machines participate in society; the interactions of people and machines are interpreted by the human participants through a form of social *bricolage*, with actions and meanings negotiated in context by the participants. The problems encountered by technicians are most fundamentally breakdowns of the interaction between customers and their machines, which may or may not include a malfunction or failure of some machine component. Diagnosis requires negotiation with both customers and machines, first to assess the breakdown and determine the problem and then to produce an acceptable solution. Understanding is not only fragile but also variable, and technicians work hard at discovering and shaping the users' understanding of the machines so that technicians and users not only can talk of machine troubles with a common understanding but also will perceive the same behaviors as constituting trouble.

While the machines are a social presence through their participation in this social world, there is also an irreducible core of the machine as technical entity. Some portion of the work is essentially technical, in that the machinery must be adjusted, replaced, or otherwise manipulated and in that specific skills and understandings are required to do these things. Technicians' interaction with the machines as technical objects colors, mandates, and sustains the interactions with customers, managers, and other technicians. The social interactions happen, in some sense, and happen in the way they happen, because the machines need to have technical things done to them. The technicians have the ability to do those things, and the rest of the relationship follows from the contract between the customers' need and the technicians' ability.

The future of work is commonly projected to contain more office work, more service work, and more technical work. Unfortunately the terms "office," "service," and "technical" admit of many definitions, and the intended meaning is only sometimes specified. Without further specification of terms, the work of technical field service is clearly modern; a closer examination, however, qualifies its modernity. Technical field service is "service" in that the work is maintaining the technological infrastructure for others. As increasing numbers of workers become dependent upon office machines, the task of keeping those machines running will necessarily grow, just as the growth of technical repair work in the past has paralleled the increase in office work and the development of office machine technology. However, the technology itself is a mix of modern and not so modern components; copiers in their present state combine electromechanical and computer technologies. Servicing copiers consequently involves being a mechanic (repairing, adjusting, and lubricating various mechanisms) and being a programmer (setting up the software correctly); the job straddles traditional categories of blue-collar and white-collar work. Although management theories claim that modern workers, both users and maintainers, will need to understand modern machines less, the technicians' job also requires learning and preserving otherwise unavailable information about the machine. The skilled use of mechanic's hand tools is combined with a detailed understanding of the machines, though neither is supposed to characterize modern industry. Ultimately, therefore, there is an irony to the use of the term "modern" to describe field service work, rooted in the very traditional nature of the technical skills involved.

While the servicing of copiers does take place in offices, the presence of these technicians and these machines in office settings is somewhat incongruous. Big copiers are far larger and noisier than other modern office machines; furthermore, copiers are dirty. The copy is produced by melting toner, composed of plastic mixed with carbon, on the paper, and there will always be a certain amount of stray toner around a machine. The technicians themselves do not quite fit inasmuch as they are not natives of the offices where they work but outsiders. Moreover, their presence indicates that something is wrong. Finally, the status of their job is uncertain. The technicians dress like office workers, and their tool boxes look like briefcases but they weigh too much. The technicians are skilled workers, but they get dirty. In modern offices virtually no one gets dirty, particularly if the job is defined as skilled. The technicians and their copiers bring more than a hint of the factory floor to the offices which are supposed to have supplanted the factories as the normal setting of modern work.

The technicians' suits are, presumably, intended to offset this suggestion of industrial processes in the office. At the time and place of this study, the corporate dress code meant that all technicians wore suits or jackets, with neckties required for men. Outsiders find it astonishing that one would dress so for such work, but the suits constitute both a claim that the wearer is a modern businessperson and a claim that the machine is sufficiently domesticated that it can be serviced by one so attired. The technicians certainly subscribe to the first claim; the second is perceived as something of a challenge.

The nature of the work of technical service as defined by the corporation is the result of contracts between corporations, negotiations between workers and management, and unilateral decisions by management in the form of service policy and the design and content of machine documentation. At the same time, the actual expert practice of technical service is necessarily an improvisation by the participants in a given situation. Each episode of machine repair is built on shared knowledge of earlier successes and failures, and the stories that the technicians tell circulate that knowledge. The stories also celebrate the technicians' mastery of the complex and sometimes obscure interaction between technicians, customers, and machines, while acknowledging the contingent and temporary nature of their success. The principal issues for the technicians in this triangular interaction are control and understanding, and one reward for achieving the two is their own identity as competent technicians. The first and foremost goal of practice, however, is *getting the job done*, and it is only by accomplishing that primary goal that practice contributes to the technicians' social identity and preserves their relations of employment.

THE FIELDWORK

The corporation described here is Xerox, as one might guess; however, this is the only place where it will be explicitly identified. I believe that very little of what I say is unique to Xerox, and I do not want to burden the observations with that identification. I would prefer that the reader think of what follows as a description and analysis of a particular job as it may be observed, and then reflect on the similarity to or difference from work and workplaces the reader knows.

It was of some advantage to me in doing this research that I had worked as a technician. In 1966, I dropped out of college and was subsequently drafted. In the army, I became a technician, repairing a variety of communications equipment. After separation from the army, I worked as a technician before, during, and after finishing my bachelor's degree. Indeed, when it became clear that my intended thesis research in Afghanistan could not be accomplished due to the political circumstances of the time (1979), I returned to being a technician and found a job at the research institution that employs me now. It was as a technician that I made my first trip to the corporation's training center to learn to repair a copier that we intended to turn into a laser printer, and this work with printers provided valuable experience in which to ground my observations of technicians for this study.

It is important to note that I was never a field service technician, fixing machines for customers in the customers' place of business. I either worked on machines that had not yet been delivered or on machines in use in-house, in the place where I worked, used by people employed by the same corporation. To some extent, this reduced the triangular relationship of service by minimizing the customer's role. This meant that my work focused more on the machines, leading me to believe that the work of service is about broken machines. It was only my work on this study that showed me how great a role the customer has in the production of a fixable problem from the situation of service.

My practical experience was both boon and curse. It was beneficial in that it made my presence in the field less obtrusive, since I needed fewer explanations. It was helpful in winning the trust of the technicians. However, it was a problem in analysis since my notes omitted things that were obvious in the field but are less so at a distance. I also found I had a tendency to regard certain phenomena as unremarkable which are not really so to outsiders. The assistance of colleagues was invaluable in calling my attention to some interesting material; that remaining in oblivion is entirely my responsibility.

The first step in my fieldwork was to attend the repair school for the principal copier serviced by the technicians I intended to observe. The main reason for doing this was to ensure that I would understand what was being done to the copier in the field. It also enabled me to participate in diagnoses in a peripheral way; I avoided more direct involvement because I wanted to know how *they* did diagnosis. A secondary reason for attending the school was to spend three weeks immersed in technician culture, getting attuned to technician stories, concerns, and practice.

After the school, I received permission from the corporation to do field visits, as the service organization refers to them. Observing the technicians involved going with them to customer sites on service calls or courtesy calls, going to the Parts Drop to pick up spare parts, eating lunch and hanging out at local restaurants with the other technicians when there was little work to do, and occasionally going to the branch, or District Office, for meetings, paperwork, or to consult with the technical specialists. All of my observations were made on the job or between calls; I did not do structured interviews. With the permission of the technicians (and their customers, where appropriate), I made audio tapes of our adventures; I also took copious field notes.

In analyzing my notes and transcripts, my goal was to discern and categorize the actions of the technicians and the topics they brought up in their conversations with each other, with their customers, or with me. The interests of the technicians as they appeared in their discourse seem to fit loosely into three broad groupings, the social, the experiential, and the existential. The social concerns pertain to the arena in which service occurs. That arena has spatial and temporal dimensions, and the technicians' concern focuses on the continuing relationships of a shifting population of technicians, customers, and machines. The distinction between technician and customer is a critical division of this population, but for technicians at work, all nontechnicians are in some category of *other*, including the corporation that employs the technicians, which is seen as alien, distant, and only sometimes an ally. The spatial dimension of the arena is defined by the territorial divisions of the service world; the temporal, by an awareness of continuity and change in the relationships among the inhabitants and by the expectation that these relationships will continue to evolve into the future. The experiential and the existential concerns are about what happens in the service arena. For my purposes, the experiential concerns are about the way things happen, the way work gets done, while the existential reflect both the technicians' sense of values and their thoughts on the nature of the work itself. The separation between these three groupings is analytic only, and a single conversational exchange may contain elements belonging to all three. My intent in distinguishing them is to separate the description of the world in which service occurs from the description of the work itself and to separate both from the description of how the technicians feel and think about the work and the world. In the real world where service is done, no such separation exists.

WHAT IS WORK?

The study of work practice is unusual; what is actually done at work is rarely examined. However, Clifford Geertz (1973) suggests that one might best understand social anthropology by looking at what social anthropologists do, and the suggestion works as well for understanding other occupations. In this study, the work itself is taken as the focus; to understand why this is uncommon, one should consider the common usage and definition of the word "work." Raymond Williams sums up both the varieties of meaning evoked by "work" and its normal focus:

> As our most general word for doing something, and for something done, its range of applications has of course been enormous. What is now most interesting is its predominant specialization to paid employment. . . . The basic sense of the word, to indicate activity and effort or achievement, has thus been modified, though unevenly and incompletely, by a definition of its imposed conditions, such as working for a wage or salary: being hired. . . .
>
> The specialization of *work* to paid employment (see UNEMPLOYMENT) is the result of the development of capitalist productive relations. To be *in work* or *out of work* was to be in a definite relationship with some other who had control of the means of productive effort. *Work* then partly shifted from the productive effort itself to the predominant social relationship. (Williams 1983, pp. 334–35)

That is, "work" is now used more to mean "being employed" than to refer either to doing or to the thing done. This somewhat modified meaning clearly suits the common use of "work" in mainstream Western industrialized societies; it is unclear how well it may be applied to the margins of those societies or to any part of other societies. However, the relationship of employment contains a presumption of doing, which may or may not be made explicit in various ways. One question to be examined through a study of work practice is how well any explicit representations of doing match what must be done to accomplish the goals of the employment.

Cato Wadel writes that social scientists borrow their definition of work from modern economists for whom work is "those activities sold on the market for a price" (Wadel 1979, p. 367). According to Wadel, this leaves the real definition of "work" to business administrators; the work for which they pay comprises just those activities that they define as necessary for production. Other activities may, in fact, be equally necessary; but since business management has not so defined them, their status as work is, at best, arguable. Wadel points out that such a definition gives a skewed perception of work as the activity of production by not including all the activities essential to production.

I would add that this concept of work seems to be focused on individual workers. The activities defined by management are those which one worker will do, and work as the relationship of employment is discussed in terms of a single worker's relationship to the corporation. I suspect the incidence of workers alone in relations of employment is quite low, and the existence of coworkers must contribute to those activities done in the name of work. Interactions among groups of workers are part of the activities which Wadel says may be necessary for the work but which are not encompassed in the normal use of the term. The fact that work is commonly done by a group of workers together is only sometimes acknowledged in the literature, and the usual presence of such a community has not entered into the definition of work.

We are left, then, with a possible conflict between work as doing, as practice, and work as activities explicitly described or prescribed in the relationship of employment. What I propose to do in this study is to leave the question of such a conflict open until after I discuss the work practice of field service technicians; then we will return to this issue. First, let us consider the question of examining practice.

An important point about the ethnographic study of work practice is that it must be done in the situation in which the work normally occurs, that is, work must be seen as situated practice, in which the context is part of the activity. My consideration of service work as situated practice derives from Lucy Suchman's work, which focuses on what plans may be and on the nature of their relationship to action, particularly as seen in the actions of persons trying to follow instructions. Her claim for the fundamentally situated nature of activity is based on the premise that human mental activity is socially and materially located: "The basic premise is twofold: first, that what traditional behavioral sciences take to be cognitive phenomena have an essential relationship to a publicly available, collaboratively organized world of artifacts and actions, and secondly, that the significance of artifacts and actions, and the methods by which their significance is conveyed, have an essential relationship to their particular, concrete circumstances" (Suchman 1987, p. 50). This means, in part, that actions, or practice, must be understood with reference to the situation of their doing.

In her analysis of action and understanding, Suchman makes the point that "in the course of situated action, representation occurs when otherwise transparent activity becomes in some way problematic" (Suchman 1987, p. 50). This point is important. Normally, the world is taken for granted, unrepresented but capable of being represented. Representation may occur in advance of encountering a situation, when one plans to do something that involves some uncertainty, or afterward, when one is attempting to understand a situation one has experienced, possibly in order to fix it. One constructs representations of the situation when it is anything other than taken-for-granted, in order to make it working and transparent again. The problematic character of the situation may be that some element of it is unwieldy, broken, unavailable, or simply that the whole situation has somehow come into question.

In such circumstances, much of situated practice is the piecing together of an understanding of the situation and of possible courses of action, and this is true of service work as well. Lévi-Strauss's concept of *bricolage* (1966) is a useful way to think of this piecing-together. The point of *bricolage* is the reflective use of what is at hand—things, understandings, facts—to accomplish a defined goal, which in the case of service work is the understanding of troubles and their solution. Understanding the situation often means defining the problem to solve, and Donald Schon's reflective practitioner is another *bricoleur* of this sort. Schon is concerned with professionals of various sorts floundering in the world because they were trained to solve problems but found themselves unable to see them. "In real-world practice, problems do not present themselves to the practitioner as givens. They must be constructed from the materials of problematic situations which are puzzling, troubling, and uncertain. In order to convert a problematic situation to a problem, a practitioner must do a certain kind of work. He must make sense of an uncertain situation that initially makes no sense" (Schon 1983, pp. 39–40). In my interpretation, the three concepts—Lévi-Strauss's *bricolage*, Schon's reflective practice, and Suchman's situated practice—all center on the interactive construction of an understanding and a basis for action in the context of the problematic situation. Such constructions are part of learning, additions to the *bricoleur's* set, and will be revisited in retrospection or when attempting to analyze new problems.

Such construction of understanding may be represented in stories or may be accomplished with stories; stories are commonly used to make sense out of ambiguous situations or to represent sense-making in earlier events. The telling of stories is situated as well; some stories only emerge in certain contexts, or emerge differently in different contexts, and those hearing a story shape it as well. Ellipsis provides a guide both to what is most interesting to those telling and hearing a story, in that all the "good bits" will be included, and to what competent members of that society are expected to know, in that such matters may freely be omitted. Stories are also told to represent the way the world is, where the tellers and listeners might fit into it, or in some cases, the way the world should be. Such stories offer the tellers an opportunity to claim starring roles in their culture or at least to counter others' accounts of who they are.

The use of stories to make sense of a situation or the world itself emphasizes their role as part of the interpretive repertoire of culture. Introducing *The Interpretation of Culture*, Geertz writes: "The concept of culture I espouse . . . is essentially a semiotic one. Believing, with Max Weber, that man is an animal suspended in webs of significance he himself has spun, I take culture to be those webs, and the analysis of it to be therefore not an experimental science in search of law but an interpretive one in search of meaning" (Geertz 1973, p. 5). The webs may be seen in daily interaction, or social discourse, and while the significance is perhaps readily understood by those whose culture it is, the ethnographer's task is to understand it from the outside. "So, there are three characteristics of ethnographic description: it is interpretive; what it is interpretive of is the flow of social discourse; and the interpreting involved consists in trying to rescue the 'said' of such dis-

course from its perishing occasions and fix it in perusable terms"
(Geertz 1973, pp. 20–21).

Of course, those of whom the ethnographer is trying to make
sense may be in the act of making sense of their situation for
themselves. This may occur, for example, in stories of diagnosis,
or it may occur in legal testimony (which may also be seen as
another sort of story):

> If adjudication, in New Haven or New Hebrides, involves represent-
> ing concrete situations in a language of specific consequence that is
> at the same time a language of general coherence, then making a
> case comes to rather more than marshaling evidence to support a
> point. It comes to describing a particular course of events and an
> overall conception of life in such a way that the credibility of each
> reinforces the credibility of the other. Any legal system that hopes to
> be viable must contrive to connect the if-then structure of existence,
> as locally imagined, and the as-therefore course of experience, as
> locally perceived, so that they seem but depth and surface versions
> of the same thing. (Geertz 1983, p. 175)

Our enterprise, then, is to make sense of the technicians' situated practice, but that practice is also a sense-making endeavor, as
is culture itself. The stories for which the technicians are famous
within the corporation are examples both of the sense they make
of their world and the process of making that sense. Our goal is to
gain an understanding of the technicians' work as they do it and
as they understand it, and to use that understanding to look at the
question of the relationship between work as it is done and work
as it is described or prescribed. In this way we will come to understand both what work is like in the triangular relationship between technicians, customers, and machines, and what value there
may be to a study of work practice.

2

Vignettes of
Work in the Field

One of the features of service work is that there is no typical day. Some situations may occur more often than others, but on any given day anything may happen or nothing may happen. Nevertheless, one can gain a general sense of technicians' work by examining vignettes of their lives; some combination of the scenes that follow or variations on these themes make up the events of most days. Three of these vignettes show interactions with people: one is a breakfast meeting of a subteam of technicians, another is a lunch scene where a technician with a problem has gone to a restaurant to meet other technicians in the hope of getting help, and the third is a courtesy call, a visit to a customer at a time when there is no service work to be done. Two focus on interactions with machines, one with an older machine whose perversities are said to be responsible for the skills of many of the best technicians and one with the new machines which are the primary responsibility of the technicians I studied. These interactions clearly show the extent to which service work is centered on the triangular relationship of technicians, customers, and machines.

To preserve the flow of events and to suggest the complexity of the tasks these technicians have mastered, I first present the vignettes with minimal explanation, recognizing that the reader will find some of the detail alien. Following each vignette comes a section of commentary which includes necessary background information about the organization, the activities, and the nature of the machines. All names have been changed to protect the identities of those collaborating in this study.

First Vignette—A Breakfast Meeting

I drive across the valley to meet the members of the CST (or subteam) for breakfast at a chain restaurant in a small city on the east side. Silicon Valley is clear this morning, so the hills are sharp in front of me and behind me. As I sit at a traffic light, I think that this is just like a suburb of industry, tract factories if you will. There are the same artistically curved roads with improbable names that one finds in normal suburbs, but these drives have five lanes in each direction, and all are full. The buildings have the family resemblance of tract houses, but these are prefabricated concrete tip-ups. They have nice lawns as well and good landscaping, but behind the landscaping are large parking lots. It goes on for miles, in a strange combination of sprawl and density; technicians may have to drive miles between calls, but every mile is on streets lined with factories much like the ones to which they go and the ones they have just left. They wear out their cars quickly in this environment, but their drivers' licenses may last no longer. There are patrol cars and radar cars, and all the police agencies use motorcycles. Homevale uses BMWs, and no one thinks of them as police bikes. The businesses along these roads are distinctive, too, being neither industrial factories nor big office complexes. The electronics industry is full of engineers, professionals and enthusiasts, who routinely work long hours and expect the world around them to keep pace, including their office machines. Since new invention is vital to the industry, security is tight in these buildings, and taking tools or parts in and out means having your bag inspected each time. These tracts of factories are interspersed with tracts of houses, blank areas on the map for technicians since few homes have copiers; shopping and other services cluster around the vestigial downtowns of the old farming communities. The soil in this valley is amazingly fertile, and there are still a few truck farms amid the tracts of houses and factories, growing their crops until the price for the land is right, or until the complaints about noise, dust, or chemicals drive the farmers out of business. The valley used to be famous for its fruit trees; chips, computers, disk drives, and printers are the harvest now.

At the restaurant, Tom, Jim, and Joan, three of the four members of the subteam, are talking about what to do with the technical specialist, Sam, who will be with them later in the week. The idea is to try to use his expert assistance to clean up machines that have been chronically troublesome, although the bad machines change from week to week. This reminds Tom that Sam had been with him the last time he had a service call on one of the machines we had visited the previous day. Then he tells Jim that the two machines at that account are still being heavily used, although less so than they had been. Jim responds that he and someone else had been to another of their problem machines not long ago for a crashing problem.[1] This provokes a general discussion, because this machine has had recurrent crashing problems, and the usually suspect components have been replaced several times. In this instance, they were replaced again, even though the symptoms that usually lead to replacement were absent. This case reminds Joan that after she replaced a set of components, Sam and Susie discovered that the new ones were defective and were causing additional problems. This leads the entire CST to tell me "how it is in the field":

Tom: We haven't had a whole lot of luck. I don't know what it is. It's not that . . . I think I can speak for the CST. We're not in a rush, not hitting and licking . . . [*which I think means doing hurry-up calls so that they only superficially fix things until the next call*]. Whether it comes in phases, we seem to be, you know those periods when you get a high call rate, a lot of call-backs, until you get all the bugs out, then they settle down for a couple of months.

Joan: They all do it all at once, though, like right now when we're getting all call-backs.

---

[1] That is, the controlling electronics lose their way and stop functioning. No parts may be broken, and no damage may be done, but the program is not running.

Tom: It seems that way, and it's not just me in particular, we're always chasing each other around. Not the following day, but maybe a week later.

Jim: Yes, and it doesn't matter whether they're 4000s or not.

Tom: Nothing we're doing differently, just a phase. Ron was saying this, too. He was almost bragging that everything he's doing runs. The following week everything hit him at once. He must have had eight calls up, all of them call-backs.

Joan: We've been going through power supplies right now. Which is weird. Didn't have the problem for a long time, then Tom and Cathy both had them, then I had an Illumination Power Supply [*problem*].

Tom: These [*current problems*] are the Low Voltage Power Supplies [*after making affirmative noises while Joan speaks*]. I wonder, have you been on any calls for new installs such that the tech rep ends up being there for the whole week for boards blowing up?

I had not, although I had been on a service call caused by an error made during installation, and further, made by someone with a reputation for not making mistakes, which had considerably amused the technician taking the call and still amuses the CST at breakfast. The conversation turns back to business mixed with personal touches. The discovery that Joan and Jim are not going to work together today produces an account of their lunch yesterday in Los Padres, which leads Tom to ask if they visited a friend who works there. Then he passes on some information from Sam, resolving questions Tom and Joan had had about the confusing configuration of one of her machines. The theme of confusion leads him to a description of the bizarre behavior of one of the machines we had visited yesterday. Sam says the machine is an engineering prototype escaped to the field, but this explanation fails to satisfy Tom because the machine has not acted this way before. Tom wants to get the components relevant to the behavior from Jim, so that he can go investigate further if he has time. While they are thinking about odd behavior, Joan brings up a machine which had been spelling its messages incorrectly; it turned out to be a leaky connector. Tom gets back to his tale, pointing out that the behavior the customer sees is perfectly normal, it is just the service interactions that are weird. This leads to a digression on how to do service interactions in the absence of instructions from the machine. They talk for a bit about all the pieces of the problem machine which they have changed and then shift to the updates Tom has done. Joan announces that she is only going to work on machines which have had the updates done, provoking some teasing from the others, while Tom tells Jim that he used the last old-style part in one of Jim's machines which we had serviced the previous day. Jim inquires if we had seen his old girlfriend's sister there, which draws more teasing from the other two. Firmly in teasing mode, Tom starts in on Joan, but the teasing turns back on Tom and traffic problems. This leads to a long discussion of the problems of Tom's new car and the difficulties of service departments both for cars and copiers, while we finish our coffee and prepare to go to work.

Tom asks where everyone is going this morning, leading to the discussion of who should do an installation. Some people are notorious for avoiding them. Jim may be transferring to another team, so maybe he should do it as a last experience with these machines, but if he is leaving, he might not care whether it ran after he left. Tom suggests a tech who is new to the machine and has been doing a lot of installations to gain experience. This leads to a round of bragging about how long machines run after various people do the installations, which quickly reaches Paul Bunyan proportions. Then they get back to business, as Tom says where he is going. Joan thought he had been there, but he had been off looking for her, thinking she needed help. She tells him what the problem was, just a switch, but it seems like it has just been changed. They figure out how long it is since the switch has been changed and how much use the machine has had in that time, which is comparatively light. Tom reminds them that machines getting light usage need special attentions. The other two joke about whether they have bothered, but then assure him that it was covered.

Getting back to the original subject, Tom tells them they have a couple of days to think of a machine which they would like to work on with Sam's expert assistance. Then he tells them who will be working the following Monday, a less than universally observed national holiday, which requires some coverage but less than a full team. He was scheduled to work but he is going to take the day off and go skiing. Joan offers to let someone else work for her, and she will go skiing too. He warns her that the cabin may be crowded, and they talk about how crowded their social life has been and how both had spent the previous weekend doing very little. Joan had bought herself a lawnmower and brought it home in a borrowed pickup truck. Her boyfriend had worried about the pickup's resemblance to a catering truck; what would the neighbors think?

And then they turn back to work. Jim asks Joan if she took a call that had been up yesterday, wondering if there was anything to do other than the installation that no one wants to do. She is doing that this morning; maybe he should call and see what is up. He could; he is going to have to cancel the two calls he took yesterday in order to get more. He forgot his book so he cannot clear them, and he is not supposed to get any more without clearing them. They tell him he is crazy; since he has done the work, he should clear them and get the credit. He can tell Work Support that he forgot his book and he will clear them in the morning. Jim says they have gotten upset at him about this before; one time he had six calls uncleared. . . . Then they joke about how to take advantage of this behavior, wishing he had taken and would cancel calls that they do not want to report. In fact, it turns out that Tom is officially assigned one of the calls Jim is planning to cancel. He had overhauled one of the machine's subsystems the last time he was there. This reminds Jim that that customer says Tom took one of their vendor badges the last time he was there, which Tom denies. None of them ever get badges there. Then we go. Tom is planning to knock off at 3:30 to beat the traffic, which amuses the others who see it as an excuse to quit early. He had done the same the previous day, but after I had left, he had gone off to try to assist Joan with a problem instead of actually quitting.

Commentary

The first thing to notice in this vignette is the sheer volume of talk, the number of topics in the technicians' daily world that are thought to be worth talking about. They talk about the work they have done, the work they are currently engaged to do, and the work they are going to do in the future. Besides work, they discuss friends, lunches, cars, and traffic cops. The world of their discourse is a rich and complex one, full of nuance, and their stories and partial stories add detail and color to particular portions of this world. The stories in this vignette have been masked in my summation of their telling. Most can be divided into stories concerned with new or unusual behaviors of the machines and personal stories. Others have a more existential focus; in the transcript of the CST telling "how it is in the field," one finds this cryptic tale of a colleague's hubris: "Ron was saying this, too. He was almost bragging that everything he's doing runs. The following week everything hit him at once. He must have had eight calls up, all of them call-backs." The unpredictability of the world expressed in this story is part of the motivation for the detailed stories of machine behavior; the technicians can never know which details will be critical on their next call. The other motivation is sheer interest in the world in which they work and the roles they play in that world.

The initials CST stand for Customer Support Team, according to the person who was manager of the team I studied. It is the organization of the service world that makes the letters CST meaningful. The national service organization of this corporation divides the United States into Eastern and Western operations; at the time of my fieldwork, the Western was divided into Los Angeles and San Francisco regions. However, the San Francisco region stretched from Salt Lake City in the east to Guam and Kwajalein in the west, and from central California north to Alaska. Regions are divided into districts, and work in the district is done from the district office. Both the district and the district office were sometimes referred to as "the branch," and I will use the terms interchangeably. The technicians in the district are divided into teams, primarily on the basis of servicing machines of similar capacity or technology. If such a team becomes too large, it is divided, and the new teams divide the district geographically. Each technician is assigned a territory, consisting of a collection of accounts more or less geographically contiguous but not perfectly so. The technicians would like more geographically compact territories, but both machine and technician populations are continually changing, and both customers and the service organization want some continuity in responsibility for a given machine, so the territories will never be as neat as the technicians would like. The technicians of a team are divided into subteams or CSTs, often because they have neighboring territories but sometimes because of shared skills. The team I studied primarily serviced a popular, new, mid-volume copier; however, the CST with whom we have just had coffee was also responsible for a family of older machines which few or none of the other technicians could service.

There are three basic categories of membership in the team: the technician who fixes the machines, the specialist, and the manager. The managers are usually former technicians and indeed may be technicians again. The rewards for being a manager are such that many technicians try the job, but the pressures are such that few last. The specialists are also promoted from the ranks of the technicians; they are in charge of the technical expertise of the team. It is their responsibility to circulate current information about new problems, fixes, or updates. They also act as consultants on difficult problems and try to work with each of the technicians to monitor their technical skill. At breakfast this day, the technicians in the subteam were deciding which of their problem machines to attack during their next day or two with the technical specialist.

The territories are an assignment of responsibility, but they do not definitively determine who fixes which machine for any given service call. The pattern of failure is too erratic for technicians to work only on their own machines. This is the motivation for their continual reporting on the status of machines, as when Tom tells Jim about the machines Tom and I had visited the day before. These were nominally Jim's machines, and so he is expected to keep track of what happens to them; but it is largely a matter of chance who will take the next call on them, and so the information is of interest to the whole CST. Knowing that the customer's pattern of use has changed will change the set of problems anticipated on a service call. In the transcribed section, Tom's comment about their chasing each other around reflects their current inability to maintain control in this situation of working on each other's machines. Jim's response that the difficult machines are not 4000s, the older machines, surrenders their best excuse, since the old machines can be expected to be difficult.

Toward the end of breakfast, Joan asks Jim if he has called to see what was up. Service calls are allocated through a computerized Work Support Center, usually referred to as Work Support. Customers call to report problems. The Work Support Center operators check to make sure the computer entry about the machine contains current information; they will also try to clarify the problem and may try to solve it if it seems to be a misunderstanding between the customer and the machine. If this strategy fails or does not seem appropriate to the situation, the operators add the call to the list waiting to be assigned to a technician. When technicians need new calls to go to, as here after breakfast, they call Work Support to check the queue of calls waiting for them and their team. When technicians finish a call, they phone Work Support to report this fact, along with numerous bookkeeping details, and then take new calls. They are supposed to take the oldest call on a machine assigned to them; in practice this is balanced against distance, what is known of the machine and the customer, and the situations of their teammates. An older call for one of their colleagues could take precedence over a call of their own, particularly if it is more convenient to where they are or where they want to be. In any case, the technicians choose calls from the queue Work Support gives them; calls are not assigned. Calls that have been waiting more than a certain period of time are supposed to have priority without regard to territory; at the time of the fieldwork, Work Support was located in the branch office, the operators were personally known to the technicians, and the priority of calls could be negotiated.

In theory, technicians cannot take new calls until they have cleared the old ones; this is what prompts Jim's remark that he will have to cancel calls, to report that there was no problem and no service call, because he does not have his book with the details necessary to clear them. Since he has done the work, and the records of calls he has done are part of his performance appraisal, his colleagues find this behavior inexplicable, especially since they have been able to get new calls in his situation.

The details of organization and territory affect the way technicians work without determining their approach. Their conversation reveals a landscape studded with specific places, individual people, and particular machines with particular problem histories. One quality of places in the landscape is freedom of access; some sites require badges, some escorts, and some deny access. Those customers bring the machine out, which means the technician never sees the machine in the environment in which it is used. The machines have a quality of being in or out of territory, and this does have an effect on the way the technicians work, but this is only one of the interesting attributes of the machine. Significant people in the landscape may be teammates, customers, or even the sister of an ex-girlfriend. The world of work is not just about broken machines, and the technicians' conversation reflects this as it flows freely from technical detail to the nature of lunch to people they used to know through the corporation or in some other context. The nominally personal and nominally professional cannot be separated in their conversation and may be substantially indistinguishable in their experience.

Second Vignette—The Older Machines

Tom and I go to a call of his on one of the older machines. This customer is notorious among the technicians for being reluctant to move up to newer machines. When we arrive at the customer's building, Tom cannot find the person in charge of the machine; indeed, there does not seem to be anyone in this part of the building. At the other end of the building, the person responsible for a different machine tells us that the person we want is out sick and everyone else is at a meeting. With some reluctance and disclaimers of responsibility, we are let into the room with the machine; we promise to lock up when we leave.

Tom asks if I am familiar with these machines; I am not, so he explains what he is checking. He examines some reject copies from the trash barrel; their partially processed condition tells him that the reported problem is not the actual problem, that the jams occur elsewhere. With that in mind, he begins to think about what the machine will require in the way of routine servicing. This machine gets very little use, so he will do some minimal cleaning and topping-up of fluids while we wait for the machine to warm up. These customers have a peculiar billing arrangement which requires a written form each time; Tom always writes in a minimum of one hour labor, believing that to be a standard policy of the corporation. The form should be signed by the person responsible for the machine, the one who is out sick, but he will sign it instead.

Then Tom starts telling me about the oddities of the machine, the old mechanisms and their quirks. Cleanliness *is* godliness in copiers, but the cleaning assembly here is a baroque construction of tubes, chains, scrapers, and augers. The whole business is located next to the oil hoses, which are made of a plastic that ages and cracks, filling the cleaner and related assemblies with a sludge of oil and dry ink. One wants to anticipate this problem and change the pieces ahead of failure. Tom had rebuilt this one last fall and shows me the long list of replaced parts. In addition to the cleaner assembly, he had also rebuilt one of the paper transports and the feeder mechanism, making a very long service call. In the three months since, it has made a trivial number of copies. As he shows me around the machine, I recognize the similarity between some components and those on machines with which I am familiar. Tom says they are similar but much messier; in fact, dirt is the hallmark of this machine. These machines always need a lot of cleaning. The optical path is complicated, too, using mirrors all around the machine. This complexity produces all sorts of image problems; when the machine was newer, you could ship it back to get the optics fixed. Now you tweak the optics until it works, sort of, but no one expects wonderful image quality from such an old machine anyway.

Tom thinks it is curious that the older machines have smaller parts budgets, smaller allowances with which to repair them, when they take the same amount of time and possibly more repairs than the newer machines. Even so, he prefers some of the older, simpler ones to the first ventures into new technology: the relay chassis that controls this machine is more reliable than the first microprocessor controller. He is really enthusiastic about the new machines, however; he says they are the best thing he has seen in fourteen years with the company.

We turn back to the conversation at breakfast. Tom tells me that most CST meetings are a chance for the members to catch up on news, tell each other about problems or what they have been doing in each other's territory, and think about how to deal with problems they face as a group. The visit from Sam is part of a new program to deal with their worst machines. The technicians make lists of their most troubled machines.[2] Then, when Sam is available, Sam and each technician in turn visit as many of the problem machines in that technician's territory as possible and really try to clean them up from top to bottom. The floater covers other calls in the technician's territory while this happens. They have been doing this for two or three months now and have already hit the worst problems. Tom says in a year you would think they would have a perfect territory, but that will never happen.

By now the machine is warmed up, so he tries it, predicting where it will jam. It does. He says he has had this problem before and we have one of two causes, one of which he dealt with last time by replacing a part. However, there is a new replacement part for it, and he will put one in if that is not what he did last time. Tom takes the covers off the machine, commenting on its filthiness and monstrous mechanisms. Looking at the vacuum mechanism, he thinks that low vacuum could be another possible cause for the problem. Looking at the relay chassis and the cycle control switches which together control the machine reminds him that a timing error could produce his symptom. However, he will pursue his original hunch first.

Tom explains to me that the mechanism he suspects has to move into position. It is driven there by a chain which is controlled by a clutch. When it reaches its position, it is latched, and the latching releases the clutch and so the drive. The latch switch is the old-fashioned variety and so it is Tom's prime suspect, but

---

[2] Not all problems are with the machines, however; some are with the customers.

the shear pins of the clutch are a possibility as well, and very easy to check. Tom uses a punch to tap one of the pins until its other end begins to emerge from its hole; he concludes it is OK. He explains that the clutch which is held by the shear pins is adjusted by this Allen screw over here, and that the shear pins are there to protect the drive gears. The drive gears are interesting, too, because they are made of Oilite, which is not supposed to need lubrication, but they get noisy so the technicians grease them anyway. He shows me the mesh of the fine teeth, the drive arrangement, and where to look for wear. "These teeth get rounded and real silvery."

Now that we have inspected it, we will watch it in operation. This machine does not have the diagnostic programs found in later machines, which permit you to exercise different parts of the mechanism for test purposes, but there are ways of inducing a dead cycle for the same effect. Tom tells me to watch as the clutch engages, the chain moves the mechanism around, the switch latches, and the clutch releases. He tells me that the clunk produced by the clutch releasing the tension in the drive chain is the most characteristic sound of this machine.

It worked this time, but intermittents are not unusual. We will exercise it with some legal-size paper and plan to change the switch even if the problem does not recur. We will also check the timing. Since it seems to be working now, Tom suggests that we go see if he has a switch in the car. He recently emptied his trunk, taking a call on someone else's machine. That machine was supposed to have been replaced with a newer one, but the replacement has been delayed while the corporation tries to talk the customer into a different machine because the one that was ordered is having "teething problems." Tom does not know what is wrong with the machine they ordered; some of his customers like them.

We walk out to the car, and while Tom rummages for the part, he tells me that he has been adjusting his trunk inventory to carry fewer parts for the older machines, reflecting their dwindling population. Some of these parts on the official inventory list almost never get used, while other parts get replaced so often that the official inventory level is far too low. He finds the switch he is looking for and writes down the part number to order replacements. The installation instructions are in his tool box. The new switch uses only two of the wires connected to the old switch. The instructions tell the technicians to cut off the redundant wire, but Tom just tapes it and tucks it out of the way. Then, if he runs out of the new switches, he can use one of the old ones and not have to break a call to get a switch.

We go back in. The fact that the machine has not failed is making Tom uneasy. We will change the switch and check the cycle control circuitry as well; the cycle control is one of the other possible causes that had occurred to him earlier. Tom tells me that although these old machines are very dirty, they are fairly simple to rebuild, and virtually the whole machine can be overhauled at the customer's site. He pulls out the instruction sheet for installing the new switch. Unlike the old switch, the new switch does not include a new mounting bracket, complicating the installation somewhat. It is also necessary to find the wire numbers to be sure of getting the right pair, a task made more difficult by years of accumulated grime.

As Tom installs the switch, he tells me about things he has learned to do or to check in the interests of greater reliability, such as screws that are apt to loosen unpredictably after years of staying put. Then he looks at the switch, wondering if he has mounted it correctly, but decides he has. He cleans the wires and starts looking them up in the schematics. With the schematics open in front of him, he starts thinking about the control circuitry that he wants to investigate next, showing me the parts involved. He thinks this circuitry is far more elaborate than it needs to be, offering far too many chances for things to go wrong. He looks for the relay which switches the signal that concerns him; when he finds it on the schematics, he notes its number and says we will check it when we are done with the switch. Then he turns to the wiring diagram to find the numbers of the wires on the switch. In these machines, the common is usually pink, which would help identify one wire, but it is not always so. He finds the two wires he wants to reuse on the schematics, so it is not necessary to find the third, the one that he will tuck out of the way.

Tom adjusts the switch position for proper actuation. The instructions say to use a shim, but you cannot actually reach in there to do so. Instead you watch it dead cycle, adjusting the switch so it actuates but the actuator does not bottom out, which would eventually destroy the switch. He adjusts it, tightens the screws, and tapes the other wire out of the way. Tom tells me that one reason for wanting to be able to go back to the old switch is that the allotted time for a service call on this machine is so short that there is no time to go get parts, so one wants to be able to use whichever switch one has. He also reminds me that the time spent servicing machines is one of the considerations in their performance appraisals; it is less important than customer satisfaction or machine reliability statistics, but it is noticed. Then Tom checks the switch adjustment again and finds that it is bottoming, so he readjusts it until it is not.

Tom mentions the breakfast meeting with the other members of the CST again, somewhat warily checking my reaction. He thinks their subteam is the closest and most supportive group of the whole team. The other technicians I rode with had showed little attachment to their subteams, so I ask him about the different groups. He tells me who is on which subteam, who has mostly old machines and difficult customers, who has all new machines. He says some of the subteams are very stubborn and individualistic, preferring to work independently, but some members of such teams come visit Tom's group from time to time. He thinks they miss the camaraderie, the joking. Tom has worked with Joan for six years now, and they are friends off the job as well. He thinks the closeness of the group surprises some of the other technicians, but it probably makes them more effective. Their manager seems to think so, anyway.

In the middle of all this, Tom mentions that now he has adjusted the switch too far out, so it does not release and the chain is very tight. This is the condition that shears clutch pins eventually. As he readjusts the switch, he adds that we do not have to worry about one of the control systems he had mentioned earlier, because it is not actually involved in this process. There is a critical relay to inspect, however; there was a retrofit in the '70s, upgrading these relays to a more durable type, so some technicians never inspect them any more. His group does, and we do. This one looks OK, which is not definitive; if he remains suspicious of it, he can swap it with another of the same sort which controls other parts of the machine to see if the problematic behavior changes too.

The conversation turns to a technician on another subteam who has just been to school and has been floating without a territory for a while, doing installations and backing up other technicians. She is going to be taking over one of the major accounts, freeing the senior technician who has had that territory to float and serve as a consultant and technical resource for the other technicians, backing up the team's technical specialist. Tom thinks she will be very good; she pays attention to detail, thinks through the problem, and will not panic when the documentation does not solve the problem. She has the skills to troubleshoot using the schematics from years working on smaller machines with just schematics for documentation. Like her, most of the technicians on her subteam are only trained on this team's new high-end machine and not on the older ones; therefore, they cannot help out the other parts of the team. They moved to this team from teams servicing other categories of machine, machines not serviced by this team. They can be and are, however, loaned back to their old teams to work on the machines they used to service.

Tom shows me a paper feed problem on the machine he is servicing. All of the components of the paper feed system are OK, and the paper is in right side up. He thinks the problem is a combination of cheap paper, a chronically dirty machine, and the low usage the machine gets, so the paper has time to sit and absorb moisture. Now, he says, he is going to drive it nuts, and he starts to exercise the machine, running single copies so the latch mechanism engages and releases each time, trying to provoke a failure. A multiple copy run would not work as well for his purpose because on such a run the latch remains engaged for the entire time. While he is exercising the machine, he writes down the figures from the billing meters, which monitor the machine's use and are reported with each service call. This machine gets very little use and could be replaced by a much smaller machine, which would be better from a service perspective. On the other hand, the machine is so old that the lease is virtually free, so there is little incentive for the customer to change machines.

Then Tom starts doing the routine checks to make sure the rest of the machine is functioning and not about to fail. Do the lamps have dark spots, a sure sign of imminent failure? Are they clean? Do they seat well in their connectors? He runs copies of a test pattern; different irregularities reveal dirt or wear in the scanning mechanism, blockages in the developer housing, or dirt on the corotrons.[3] Everything seems OK, so he will top up the oil, but he knows they do not have any in stock. He will have to get some from the person who let us in from the other end of the building, who will not like it because it will never be repaid. In theory, both ends of the building are part of the same corporation, but one end never buys supplies, saving their budget, while the other resents having their supplies used for the other's machine, depleting their own supplies budget. Tom could turn off the machine without oil until the persons responsible get the necessary supplies, and he could even charge them for turning it off, but that is an extreme measure that he is not now willing to take. He will write on the oil bottle that it was borrowed from the other division, and in the paperwork he leaves reporting the service call, he will include a note that they must buy supplies.

Tom is finishing up now. He writes down the list of parts he replaced and notes those that need to be reordered because his trunk inventory has reached its threshold. Then he makes the final copies for his copy quality check; with this machine, you can hear each function in sequence, and he calls them off as the significant noises occur. I ask him if he can do this with the more modern machine as well; he can for one subsystem, but most of the machine is too quiet. This machine displays poor copy quality, which he attributes to the use of supplies from other manufacturers. He finishes writing up his own record book entries and the documents he will leave for the person responsible for the machine. Then he calls Work Support. He has to report a machine we serviced the day before as well as this call, then he asks for messages and new calls. One new job is an installation, which he rejects, claiming it is not in his territory. It used to be his territory, but no

---

[3] The corotron and variants such as dicorotrons (or dicors) and scorotrons are devices to charge the photoreceptor electrically by creating a corona, an ionized field which forms around a high-voltage wire.

more. He services other machines at other locations for that customer, but they should not be his responsibility either. He gets the phone number where he should be able to reach Joan and takes a call to go visit a very-high-volume machine which he usually has to service twice a week.

Then we lock up, leaving keys and papers on the desk of the person responsible for the machine. We go to the other end of the building and explain to the person who let us in that we are done and that we have locked up. Tom also explains that he had to borrow more oil and that he left a note for the others to stock up. Then he thanks her and we leave.

Commentary

This vignette illustrates some of the differences found in the older machines on which so many of these technicians learned their craft. Tom's assertion that dirt is the hallmark of this machine sets the stage; all copiers are dirty in comparison with other office machines, but the older ones are much more so. In some ways it is easier to diagnose this machine than newer ones: when Tom listens to the machine, he can distinguish the sounds of each step in its operation. Should it fail, he knows what it was doing at the time. This is only occasionally possible with the newer machines, which are both quieter, so there is less to hear, and faster, so it is more difficult to distinguish the operations, some of which happen simultaneously.

The amount of technical detail included in this narrative emphasizes the amount of detailed information that is necessary for the technicians' daily rounds. Consider the autonomous nature of their work. Each technician travels with tools, parts, manuals, and records and is essentially independent, except that it is never possible to have a complete set of parts or information, so the technicians depend on each other to fill these gaps. Even a reasonably complete set of parts may be depleted by a service call on a machine that needs many parts replaced.

Some of the necessary information is relatively esoteric. In the brief allusion to a paper feed problem, Tom made the point that the paper was in right side up. Cut paper has a curl in its long dimension, and when it is loaded in a copier, one side or the other of this curl must be up. A copier will misfeed if the wrong side is up, but the right side differs from machine to machine. This somewhat obscure fact is not just something the technicians must know but something they must teach the customers who load the machine. Otherwise they will have numerous service calls for a problem which is not actually caused by something being wrong with the machine.

The course of Tom's diagnosis of the cause of the problem is worth tracing. When we walk in, he has been told to expect a certain type of problem. The discarded copies in the trash are enough for him to conclude that the report is inaccurate. As soon as he reproduces the problem, that is, as soon as the machine jams for him, he has two candidates for the cause. As he opens the machine and looks at it, thinking with the aid of the machine in front of him for reference, two more possibilities occur to him. Note that each of the four possible causes which Tom has considered produces the same symptom; each produces a jam that leaves the paper in exactly the same place in the paper path, at the same time and in the same condition, as the others. All four would appear to the user to be the same problem because of that common symptom. Note, too, that two of the possible causes are not hard failures in the sense of something definitively broken; instead, they are caused by systems still functioning but not functioning well enough for the other systems with which they interact.

As the call progresses, the problem goes away; the machine no longer jams. One of Tom's candidates is eliminated when he checks the shear pin, and another drops out of candidacy when he realizes it is not actually involved in the process. He addresses one of the remaining two causes, the defective switch, even though there is now no apparent problem with the machine. Since the problem was there when first tested, and nothing has been done to eliminate it, its disappearance is not to be trusted. The fourth candidate is never addressed.

There are several points to notice about Tom's work practice. In exercising the machine, Tom induces a dead cycle; this means that a part of the mechanism is operating continuously but without effect, because the mechanisms with which it interacts are rendered inoperative. Seeing the system in action is useful for troubleshooting; dead cycles can focus the action in the area of interest without wasting time performing the rest of the copier operation. In his search for the wires on the schematics, Tom is trying to find the simplest way to determine which of the three wires on the old switch is unnecessary for the new switch. Sometimes the wires have been color-coded; knowing which wire was ground, or common, by color would simplify his task because he would then need to find only one other wire and its number on the schematic. It is worth emphasizing the point that he does not install these switches as directed; he does not cut the wire because he may later need to change back to the earlier type of switch. Finally, the fact that he actually has schematics of the controller chassis is noteworthy; schematics of subassemblies are no longer issued to the technicians.[4] All they receive now are schematics of the connections between subassemblies; the premise is that if the problem is not between the subassemblies, one should replace subassemblies. It is also true that the older technology was more accessible. In this case Tom is working with a relay chassis; the wiring of the chassis can be checked, and some relays can be checked by swapping them with a different relay of the same type. This sort of diagnosis cannot be done with solid-state controllers soldered to printed circuit boards.

The most interesting narrative thread in this vignette is the story of this particular machine, and the machine is discussed both as an individual and as a representative of its type. The individual history includes what Tom has done to it, replacing the oil hoses and rebuilding the cleaning assembly, with an accompanying narrative of the disaster that can ensue if this is not done in time. The type history includes organizational arrangements which used to exist to cope with the optic woes that afflict the machine, and modifications which occurred in the 1970s and their impact on service today. Tom tells me about the idiosyncrasies of its design, such as the need to lubricate gears which are not supposed to need grease. The story of this machine in this

---

[4] Actually, it is not clear whether these were issued to Tom or whether he had obtained them through unorthodox channels.

place includes the fact that it gets very little use, so that its work could be done by a newer, smaller machine (which someone else would service), and the fact that lease charges on old machines are minimal, so this replacement is not likely to happen. The social melodrama of supplies is part of this story, too. Those with custody of the machine do not buy supplies for it but borrow from their neighbors, and so, if the machine needs supplies during a service call, the technicians must also borrow from the neighbors. This complication is part of the story of this machine, and everything in the story is part of the job of servicing the machine as well.

There are other narratives in this vignette. There is one about another technician, which reveals the changing organization of the technicians. The team evolves, recruiting people from other teams, assigning people to territories, freeing other people to float, to take excess calls, fill in for others, or serve as consultants to the team. Seen from the individual technician's perspective, this is a story about the way one's career may develop, moving from simpler machines to more complex machines, possibly becoming sufficiently skilled to be a technical specialist, a consultant. Technicians mostly remain technicians, however; there are few career opportunities within the service world in which one becomes something other than a technician. This narrative also shows some of the work practices technicians value: Tom's favorable assessment includes thoughtfulness, attention to detail, freedom from panic, and resourcefulness when the documentation does not provide the answers. One can draw from this narrative both a sensible construction of the technicians' organization and a model of how one should operate as a member of that organization.

Finally, there is talk about talk; twice in this vignette Tom returns to the topic of the breakfast meeting we had attended.[5] The first time Tom tells me that the meeting was to share information and that the visits of the technical specialist are intended to help them fix their most troublesome machines. The most interesting part of this statement is the existential despair expressed in his observation that the territories will never be trouble-free, no matter how often the technical specialist comes to help. The second

---

[5] This was described in the first vignette.

time he brings up the meeting is more problematic for him; their CST has developed personal bonds in addition to their professional association, including friendship outside the job. His narration reveals a division among technicians over the propriety of such bonds, and while he clearly values them, he is wary of revealing them to me.

THIRD VIGNETTE—A COURTESY CALL

Things are quiet, very quiet. There are no calls up at all, or at least none that Bob would take. He decides that we should pay a courtesy call on one of his accounts which has several machines. Such an account is important on its own; in this case there is an additional motive in that Bob and John, the team manager, were supposed to take the account manager to lunch yesterday, and both completely forgot. We drive over to Integrated Circuit Corporation, and I sign in at the desk; then we go to the account manager's office. Bob begins by apologizing for forgetting lunch the previous day; fortunately, the account manager forgot as well. The three of us then walk around the building, visiting each of the machines. Bob checks how much use each machine has had since his last visit and whether there are any significant problems appearing in the error logs. We talk about the idiosyncrasies of each machine and its location.

In their copying center, Bob points out that one of his machines has had one of its peripheral features disabled in order to discourage users and shift some of the work to an older machine which has the same feature and is designed for a much higher volume but is harder to use. While we are on the subject of controlling users, Bob also tells me that there have been fights between the engineers who work in the building and the "key op" responsible for another machine, which is in a "walk-up environment," that is, available for casual use. Bob and the account manager agree that engineers believe they have a right to fiddle with any machine they encounter, whether they know anything about it or not.

We go outside and talk on the sidewalk for a bit. Bob explains that he will soon be assigned to a new territory and that a different technician will service the machines at Integrated Circuit Corporation. The account manager is not happy about this and tells us that Integrated Circuit used to use the machines of a competing corporation. The technicians maintaining those machines were replaced by newer technicians who seemed incapable of fixing the machines properly or quickly, so Integrated Circuit switched machines. They have been happy with Bob's service and therefore with the machines, but the account manager is uneasy at the prospect of a different technician taking over.

The conversation then turns to personal matters, trips each is planning and other upcoming events. They commiserate on the hardships of being a single parent in Silicon Valley, particularly the fact that day care gives an unyielding and invariant deadline to the end of the day and at a time when most Silicon Valley enterprises are still bustling. On that note, we say good-bye and head for the branch office so Bob can be sure to get to the day care center on time.

Commentary

Service work is inconsistent in its pacing; sometimes there is nothing to do, and sometimes the technicians cannot keep up. If there are very few calls up, that is, in the queue at Work Support, technicians do not have to take one if they have other things to do. When technicians call the Work Support Center to check for waiting service calls, they may check just the calls on machines assigned to them or, more commonly, the calls for their subteam, the CST. They will probably take any such calls unless the pace has been extremely slow and they know that their colleagues are also waiting for something to do. If there are no calls for the CST, they may check the calls for the entire team. However, if there are no calls for them and no calls or no urgent calls for their CST, there are other work activities which they may do, such as courtesy calls, inventory of their spare parts, or updating their documentation. At this point there were no calls for Bob and no calls for anyone else that he felt obligated to accept.

Courtesy calls are not an official activity; they are not part of what the corporation says is involved in the job. However, when there are no service calls to take, the technicians think it is a good idea to visit important customers to see if they are content with their service or if there are any problems they have not reported, and the technicians' immediate management encourages them in this. Courtesy calls can be used to visit machines that the technician believes will soon need service to try to anticipate the call. They can also be used to hide "band-aid activities" intended to keep the machine running for a while without actually fixing its real problems, but that certainly was not the case here.

The machines Bob services record events in the machine's history automatically, both routine events, such as revolutions of the photoreceptor, and errors. The errors recorded are actual jams or failures as well as timing measurements that are out of specification; they are detected by an array of sensors distributed in the different parts of the machine. Reading these logs during a courtesy call tells the technicians what has been happening with the machine, when to expect its next service, and what sort of problems to be prepared to fix at the time.

One machine visited in this vignette had had a feature disabled to encourage people to use a different machine. This is just one way of managing the users; the interesting point here is that in this case the action was requested by the customer. This example leads to the story about the key operator defending the machine from the engineers; the story is told by the technician but supported by the customer's account manager.[6] Although the machines are at the customer site for the customers' purposes, the technicians see the machines as theirs and do not want most customers doing anything to the machine other than replenishing consumable supplies. The technicians might have told the key operators, the persons at the customer site immediately responsible for the machine, to keep other users out of the machine, and in this story the key op is willing to try.[7] This story reflects the tech-

---

[6] "Account manager" is almost certainly *not* the title used within the customer organization; it refers to the person's relationship to the corporation employing the technicians.

[7] Key ops are taught how to replenish supplies and how to clear paper jams, for example; they have license to do some things to the machine. The term "key

nicians' proprietary attitude toward the machines; at this site, the key operators and the account manager seem to identify with the technicians' perspective and perhaps share with them both the "ownership" of the machine and its defense against the unwashed heathen engineers.[8]

The technicians' opinion is that uninformed fiddling can do no good and could do real harm to the machine; moreover, it interferes with the natural course of events. That is, there is some predictability to the way the machine ages, the way components wear out, and the way adjustments shift between service calls. There is no way to predict what someone who does not know the machine might do to it. In some sense, the technicians never know what they will find on a service call, but there are some things they expect. Uninformed user intervention introduces an arbitrary and random factor which effectively negates some of what the technician has learned through experience, so it is not surprising that they ask the key operators to try to prevent this. The key operators, on the other hand, work for the same corporation as the engineers and are generally of lower status. Their efforts to help the technicians by keeping the engineers out of the machine are inevitably compromised by the weakness of their position in the corporation. The operators' and technicians' best hope is to enlist the sympathy of the account manager, who may have enough status to curb the engineers' explorations; it seems that this strategy may have succeeded here.

The account manager's story is undoubtedly a warning: they have changed vendors once and can do so again. These machines are usually leased, not purchased, and cancellation is the customer's ultimate recourse if dissatisfied with service. The account manager is happy with the level of service that Bob is providing. If Bob has to be replaced, the new technician should be as good. Moreover, the new technician also needs to be perceived as being as good as Bob. There is no way of knowing whether service on

op" actually derives from the fact that they are given keys which permit them to open the normally locked doors of the machines, allowing them access to the inner workings.

[8] The engineers are, of course, more educated and more highly paid than any of the "owners" except maybe the account manager, and of a higher class as well.

the other brand of machines was really inadequate; the fact that the customers thought so was enough.

The personal conversation is not uncommon. Since Bob is responsible for these machines, he sees the account manager often. The effect is to establish some common ground apart from the work context, and it is not surprising in Silicon Valley to hear of a conflict between family and work life.

Alice has a problem: Her machine reports a self-test error, but she does not quite believe it. So many of the parts of the control system in this particular machine have failed that she suspects there is some other problem that is producing the failures. She is unwilling to accept the coincidence that so many failures could be independent of each other. We are going to lunch at a restaurant where many of her colleagues eat to try to persuade Fred, the most experienced of them, to go look at the machine with her. If this fails, she will try to get the team technical specialist to look at it. She makes copies of the information from the error log and service log to take with us. In the parking lot, she recognizes the cars of her colleagues, including the one she wants to recruit to help. She says he will probably greet her as "trouble-come-to-call."

Instead, after everyone says hello, he asks her if she had been working on a particular part of the machine he is now fixing. She does not think so, which is fortunate because he wants to give someone a hard time about the state of the machine. Alice shows him the copies of the logs, and they fence for a while about why or whether he should be interested. When she succeeds in getting him and the others interested, they listen to the list of problems the machine has had and begin to talk about noise problems or communications problems; they also suggest, jokingly, that she swap the machine. She repeats that she wants help, that she does not understand this series of problems, and Fred tells her she has to fix it. He looks at the logs and tells her ways to approach the problem as a noise issue; he also tells her she cannot ignore the error code. They figure out which board the error code is actually indicating, where it is in the machine.

Fred comments again on the number of failures and asks for her copy of the manual, but she has already given it to him. She repeats that she wants help; he repeats that he is not going, but he will tell her how to approach the problem. He shows her where it tells how to check the communications lines, and they all laugh at one of the suggested remedies for persistence of the fault, a suggestion to swap all the boards in the machine, one at a time. Fred asks her again about the error code; she tells him, adding that it is persistent, and the machine will not do anything. He says in that case she should be able to find it right away with the procedure associated with the error code and why is she bothering him? She reminds him again of the number of previous failures and the number of modules replaced. He asks about a specific one as she recites the list, and yes, she has replaced that one too. One of the other technicians points out that that only means that the part is new, not that it is necessarily functional, and all the others agree. Alice repeats that she thinks there is something about the machine that is causing the failure of all these components; the other technicians all tell her just to fix the problem.

Alice reminds Fred that some of the components have been replaced twice in recent months. Fred starts to tell her about running the noise test, and then says she probably cannot do it if the machine will not run at all. Alice doubts that the machine will run enough to do the noise test, although she did manage to read the error log; if she can do that, she may be able to use some of the other diagnostic programs. He starts to tell her about testing the communications lines as she asks who is going to come help her. He tells her nobody, and goes on, as other technicians urge her to pay attention to him and she replies that she always does. She ends up reciting the procedure with him. He asks why she is giving him so much trouble if she knows all this; she says she is doing it because she is not going to get the help she wants. Fred continues that after she checks the communications lines, she should follow the procedures the book specifies for that error code. Another technician asks Fred if the manual specifies testing the communications lines the way Fred just described; the book only says to check them but Fred prefers more rigorous testing.

Fred again tells Alice to follow the error code procedure in the book if the problem is still there when she is done testing the communications lines. The error code might be correct, but given the history of the machine, it is worth checking the communications lines first. She agrees but says now she is going to go see if the team's technical specialist will come help her. Fred tells her he is unavailable; he is working with another technician on an installation. She wants to know who she can get to help, then. Fred says she does not need help, but she repeats that she wants it. Fred says she can do it, she knows enough about this machine. She responds that she knows too much about this machine. He says she should follow the procedures he has told her, that she has got to do it herself.

Fred tells her that there is another component that she needs to change according to his interpretation of the logs. She asks why, since she has only found one machine that appeared to require a new one. Fred does not explain but simply tells her to change it, that it is a lower priority than getting the machine running again, but that it should be done. Then Fred asks her about the output module. She explains that one of its functions has been disabled, but that people still try to use it despite signs all over the machine saying it does not work. That is why it shows up so often in the error log. Fred says he is not interested in that but in the fact that it is there. It is notoriously easy to pinch the communications line to that output module in the installation process, sometimes cutting it in half. He says she should check that one particularly carefully. So, he says, check the communications lines; if those are OK, try the diagnostic procedures for the error code. If the problem persists beyond that, page him.

Fred then tells her he is working on one of her machines that is using bad toner. Alice knew that and had told the key op that it was a problem. The key op had not known that it was a competitor's brand; the person on the phone had given the impression that they represented the company that made the machine. Fred says he has taken a bottle of the toner to have it analyzed; he will give them a bottle in its place. Lunch breaks up, as the various technicians start heading back to work or calling in for their next jobs. Alice is not feeling inspired; she tells me she appreciates Fred's suggestions but she does not have any confidence in them.

She knows his suggestions are based on experience, that they are things he has found to do to shortcut the diagnostic procedures. She says he wants her to check the communications lines because there is an entry for that in the error log. Following the diagnostic procedure for the displayed error code is also a logical thing to do, but she is worried. She has replaced many boards already, and she is afraid that there may be some other problem that is destroying them. She is worrying about the damage that might be done while she puts in yet another board. She has already spent enough money on this machine. However, there is nothing else to do, so we head off to do what we can.

Commentary

There are many inexpensive restaurants scattered around Silicon Valley; some of them have been adopted by the technicians as places to hang out. This restaurant is one, and there is a definite group of people that Alice expects to encounter there. They eat there, they have meetings there, they use the pay phones to call Work Support. Their managers may look for them there if they want to chat away from the branch, or District Office. In fact, technicians do not hang out at the branch; the only official place used as a hangout is the Parts Drop, which in this case is the warehouse that receives and distributes parts to the technicians. Parts Drops may be as little as a hired storage unit where a delivery service can drop off parts ordered and pick up parts being returned, but in most cases it is a place where technicians can relax, away from customers and most managers.

The reason Alice is so anxious for someone to come help her is that she has no faith in the diagnostics on this problem.[9] She knows they will tell her to change the board, but she believes there is a deeper problem which is causing all these boards to fail, and she knows that the diagnostics do not consider that possibility. She wants someone to come look at the machine with her because she does not trust herself to have noticed everything that would be

[9] The problem of *gnosis*. She can generate more questions, good questions, valid questions, than she has resources to answer, if indeed there are answers available in her world.

significant to a more skilled technician. She is also worrying about her parts budget, which is one element in her performance appraisal. Replacing boards is expensive.

I will argue that diagnosis is a narrative process, that the process of diagnosis is the process of producing an account of the troubled state of the machine. Alice's problem here is that the story produced by the diagnostics is inadequate, in her opinion, because it does not account for all the other boards that have failed. It is not certain that they do belong in one story, but the pattern of similar failures is suggestive. Her task at lunch is to tell this larger story in such a way that Fred will see the parts of it that do not cohere as loose ends to be tucked in, rather than as signs that her account includes disparate elements which do not belong together. Fred's response, ultimately, is that she must pursue the account at the level of the problem as reported and see if there remains behavior not accounted for. Such behavior might indicate the need for a more inclusive story; its absence would suggest that the account, along with the problem, is complete in itself, even if it is unpleasantly familiar.

In the meantime, all the technicians at lunch have a turn at creating a coherent narrative from the pieces that Alice has brought. Some of their stories would integrate her succession of problems into one story, and others would keep them separate. They ask each other about those parts of each other's stories that do not fit with their understanding of the machine. Ultimately, they defer to Fred's assessment that Alice must first try to conclude the narrative at the level of the immediate problem rather than begin on the more comprehensive narrative that she fears is necessary. Alice concedes that she cannot create by herself an account which covers the history of control system problems and that Fred will not help her do so. It is clear, however, that closure of this current problem in isolation will appear to her to be a suspiciously pat ending. Curiously, there is also a narrative failure to observe at this lunch. Fred tells Alice she should change a component not otherwise involved in the discussion or the problems. She asks why, since in her experience only one machine has ever needed to have this component changed, and he does not respond, he does not tell her the story. Their models differ, and he would have her follow his without providing the information to make doing so sensible.

The suggestion of trading machines is not a serious one: it is a technician's last recourse, and very expensive. Machines in this class have high prices, and the corporation does not readily give them up. Trading does happen, but the necessary levels of management are not lightly persuaded, and one may use up inordinate numbers of accumulated favors in the process. Moreover, trading machines is the ultimate abandonment of skill, since it involves claiming that one cannot cope with the machine. One necessarily loses status in the process, unless the machine is so obviously perverse that one can reasonably make the argument that no technician could cope. Technicians who have machines that have been seriously neglected may suggest swapping them rather than do all the work of bringing them up to date, cleaning them thoroughly, and replacing all the worn parts, but it seems unlikely that this suggestion is ever accepted.

The fact that the technicians cannot readily remember where a board is in the machine by the name of the board is not surprising; when technicians are not actually at a machine, they seem to have difficulty thinking about it in such detail. The fact that the documentation suggests changing all the boards is typical of the lack of realism that technicians find in the documentation. No technician would ever have all the boards at one time; the cost is prohibitive. Suggestions like this do nothing to enhance the reputation of the documentation or its designers.

Fred is trying to restore Alice's faith in herself and in her resources. From his perspective, she has to believe she can take care of things on her own, or she will be a drain on the team. He is promising that she will get the support that she needs, particularly when the known remedies really do not work, but she has to try them first. Ultimately she does not have any choice. From her perspective, this may seem brutal, but from the team's perspective, it probably seems necessary.

The toner story is familiar to all of them, and the issue of bad toner is one that plagues the technicians. The corporation sells copier supplies, but by law cannot require its customers to use them. The technicians can sell toner, and some do, but most find the commission paid to be so trivial as not to be worth the effort. Some of the competitive toner is of good quality and is sold honestly; the subject of the story here is a racket called the "toner-phoners." In this racket, someone calls the key operator and offers what sounds like a chance to buy supplies from the copier corporation at what the person on the phone says is a very good price. In truth, the prices are generally high and the toner bad, but when the shipment arrives on the dock, the customer is stuck. This also hurts the copier corporation, even though the toner is not theirs, because it was sold using their name and angers their customers. The technicians, as the part of the corporation closest to the customers, hear about this all the time. They may, in fact, be the ones to tell the customer that not only did they pay too much for something which was not what they thought but it does not work, either. The technicians are not enthused about this.

Fifth Vignette—The Routine Service Call

Frank and I are heading for the first call of the day, but he is having trouble finding the building. This is not his customer, his account, his machine, it is out of his territory, so he does not know which building it is. He grumbles that Californians do not put numbers on the buildings, and if you ask directions, they cannot give directions either. The users reported problems with the RDH,[10] which does not surprise him. No one has worked on the machine in a month and a half, and things will get dusty. We find the location, and I have to sign in as a visitor. Frank has badges for most of the businesses he visits regularly and even some he does not, like this one.

Frank finds the person in charge of the machine to learn about the problem in more detail. This person is a manager and cannot speak from direct experience, so Frank asks to talk to someone who has encountered the problem. He would also like to see the originals that were being copied when the problem appeared. While the manager goes to find the users, Frank brings up the

---

[10] The Recirculating Document Handler, an input device, automatically positions a stack of originals one by one on the glass for copying.

machine error logs and begins writing them down, commenting that he habitually does this first to make sure he does not erase them before they are recorded. He says they show a lot of errors in Tray 1. While he is writing, he explains that he wants to hear from the user who actually had a problem with the machine in order to get the most accurate possible description. Sometimes it can be hard to locate the particular person, but he finds that secondhand or thirdhand descriptions of the problem can distort things. Users may think of parts and functions of the machine in terms different from those of the technicians and may, in fact, use terms that the technicians employ for quite different parts or functions of the same machine. This log has a lot of errors; the service log indicates that Bob has been here since the last service call, but he did not write down the contents of the error log at the time. Frank wonders if he looked at it.

Two users show up; they do not have their originals, but they can assure him that they were flat, new, unwrinkled, and had never been stapled. They show Frank where the originals would get caught and tell him that when this happened there would be two pieces of paper on the glass. This happened when copying two-sided originals. This description tells Frank what the problem is, that there is too much play in the reversing roll, and he confirms it by wiggling the roll.[11] The steel D-shaft on which the roll is mounted is wobbling loose in the plastic gear that drives it. Timing is critical in this part of the machine, he says, given the speed at which it is running.

Now that he knows what occasioned the service call, he turns back to the machine logs. He says all the other entries are about right for the length of time since the last service call, with only Tray 1 and elevator faults appearing to be real problems.[12] It may be some interaction of the two that produces both entries. He starts to write up his shopping list for parts to replace. When he first comes in on a service call, he brings no parts and only some of his tools; after he has some idea of what is needed, he goes back to his car to get all of it. The photoreceptor, for example, has

[11] The reversing roll is part of the Recirculating Document Handler; it turns the paper over to copy the other side in duplex copying.
[12] The elevator raises a stack of paper into position to feed.

129,000 images on it, 45,000 revolutions; it will be at the head of the list.[13] People who want to use the machine show up; Frank tells them it will not be available for a couple of hours.

He looks at the retard roll on Tray 1; it is slick with fuser oil. Either the customer does not understand how this machine does two-sided copies, he says, or they are using the wrong transparency stock but trying it a second time due to its expense. Either way, they are putting output back in the tray to feed again, but once the paper or transparencies have been through the machine, they have fuser oil on them. The retard roll wipes off the oil; once oily, it no longer retards, so the tray feeds two sheets of paper at once, and there you are: a Tray 1 fault. Tray 2 is full of prepunched paper. Frank says this is another source of trouble, since the punched-out pieces are often mixed in with the paper and can fall off anywhere in the machine, causing a variety of problems.

The technician whose territory this is will be taking over Frank's territory, which is all the machines leased by a single customer, a major account. (Frank is going to the Engineering Products Team.) Frank says Bob will have to be a quick learner: all the machines at the major account run a very high volume, and problems show up quickly.

Frank looks at the switches that detect elevator position as a potential cause of the elevator faults, especially checking how tight the mounting screws are. He jokes that manufacturing tightens all screws until something breaks and then backs off a quarter turn. Most screws on the machine are overtightened; when screws mounting switches are too tight, the switch may bind. If the switch is not actually broken, loosening the screws and actuating the switch a few times may eliminate elevator faults. He has done this to all the machines at his account, and any time he takes a call in another's territory, if a switch is suspect, he will loosen the screws, tap the switch, and actuate it a few times. Often this cures the problem.

Frank puts a label on the feed head telling the users to press it

[13] A photoreceptor is a photoelectric device which will hold charge in the dark and discharge when exposed to light, making it possible to project an image on the photoreceptor and reproduce the image in patterns of remaining charge. This makes it possible to develop and transfer the image to paper and is the key to xerography.

down (after they have been told by another label to unlatch it to clear paper). He thinks this machine needs more labels for casual users. Then he looks around the machine. He makes a note to check their supplies of fuser oil, which they keep separate from the machine. He inspects the dicorotrons, commenting that he does not think you need safety glasses to do this if you wear glasses, but that anyone without glasses should wear the issued safety glasses. He knows one technician who does so; most of the others do not.

As we go out to get his parts, we talk. I ask about the fix for the loose gear. He will fix it by shimming the shaft with a piece of plastic banding material. There will be a retrofit, a new gear with a metal insert, but until it arrives in the field, this works very well. In fact, it makes it very difficult to remove the gear. Frank tells me he went to school for this machine eleven months ago. His account got their machines three months later, he installed them all, and he has fixed them ever since. The managers at the account are happy. The machines replaced a competitor's brand, and the volume of copying has been about twice what the managers expected. The users, on the other hand, have been unhappy, because they feel that they see the machines being fixed all the time, and they do not appreciate that the use of the machines is double that of the old ones. These same users, however, cannot be bothered to learn how they are supposed to use the machines. They claim they do not have time, he says. They seem to have plenty of time to stand around and bitch, though. He has tried running seminars to train key people in each office, but they have not had good attendance. Frank, his manager, and their cohorts in sales and customer support have a new series planned, and the managers at the account seem to have realized how much this is costing them in downtime because they say they will be pushing people to attend. There are things to learn with a machine this complex, such as what transparencies to use, how to copy cover-up without jamming, how to do duplex. Doing duplex improperly causes the retard roll problems he showed me on this machine; Frank carries twenty retard roll kits, while the prescribed level is one.

He finds this very frustrating, and the customers do not respond to his frustration. He thinks you probably have to get their

attention by billing for nuisance calls and persuading the customer to pass on the billing to the department where the machine
is located. Four or five thousand dollars of extraneous repair bills
might get their attention, and then they would come to class. The
customer corporation could do this, the billing structure permits
it, but they will not. So he is moving on and Bob inherits it.
Frank does not think any technician should stay in a high-volume,
single-account territory for more than six months. The pressure
builds as the frustration mounts, and he feels that he overreacts to
his users now. These machines are getting a lot of use, four to
eight times the national average volume, and this creates problems. There is no other machine to use in a high-volume walk-up
situation. The customer had wanted their people to take big jobs
to the reproduction department, which has machines designed for
really high volumes, not just copiers but a printing press and a
bindery, but the reproduction department is busy too. Their copying needs have mushroomed, and Frank does not see any reason
to think anything about that is going to change.

While we have been having this conversation, we walk out to
his van, find the parts, and go back. Frank removes the photoreceptor module and cleans the inside of the machine, vacuuming
everything carefully. He removes the sump which catches all the
toner collected by the cleaner assembly; he will take it away rather
than put it in the trash. If someone curious opens the sump, it
could make an awful mess. He holds on to the chassis of the
machine to ground himself while vacuuming; this is a habit left
from older machines where vacuuming the debris could generate
enormous static arcs and shock the technician. He cleans the
screens for the cooling fans; clogging there causes overheating,
which in turn can cause a wide variety of other problems. He is
also looking for all the places where the debris from the punched
paper could be and cleaning that up.

After cleaning the inside of the machine, Frank puts the new
photoreceptor in and closes up that part of the machine. He
changes the retard rolls on the paper trays, replacing the oily ones
with fresh ones. He fills the fuser oil tank. Then he turns to the
RDH which caused the call. Removing the gear is easy; putting it
back on with the shim in place requires a hammer. After he gets it

all back together, he starts exercising the machine, trying each of its different functions. While he does so, we talk about how technicians learn to do all this.

He thinks they learn enough in the school to survive their first few calls, long enough to see how the machines behave in the real world. The real problems like this gear are not covered by the official procedures, and new technicians need to learn to find them on their own. He finds the procedures useful when dealing with problems that do not occur very often; they are also useful when he is stuck. Some of the procedures do not work, some of them are circular; if the official procedure does not seem to be getting anywhere, a technician should call the technical specialist for help. Quite apart from diagnosis and repair, the machine just needs to be cleaned. Users will think the machine is broken when a technician is just cleaning it. Alice has a line ready when the customers come up and say, "Is that thing broke again?" She says, "No, it's just getting a bath."

Frank starts exercising the RDH, using fresh copies as originals; the oil on the fresh copies makes it harder for the RDH. This part of the machine does so much that there are lots of parts to wear. He thinks this subassembly is more complex than the whole of some earlier machines. The machine jams a couple of times while he is running it; he comments that these jams will be reported as RDH jams because there will be paper left in the RDH that needs to be cleared, even though the actual jam occurred elsewhere in the machine. Frank finishes his RDH checks by trying the interrupt facility, making sure the machine does the right thing.[14]

Frank looks at the finished copies, checking the functions of the output module. He inspects the staples: are the legs even, are they straight, how wide is the gap between the ends of the legs? He says with staplers you are much more likely to get wandering leg, a leg askew, than too short a leg. Some of the things he looks for he learned in the field; they were not taught in school. I ask if there is any official way to report things a technician learns like this; he does not think so. They circulate informally among the technicians. He would tell someone fresh out of school what to look for,

---

[14] The interrupt facility permits users to interrupt a programmed job and do another job without losing the programming for the first job.

but they might or might not remember. Not all of these machines have staplers; someone whose territory does not have any or many staplers may not remember those details.

The weekly team meeting is supposed to help these bits of information circulate. Frank says that John, the team manager, always starts the meeting late because the team members are sitting there talking about all their problems with the machines, teaching each other what they have learned. This only works with their team. Sometimes they have meetings with other teams that service some of these machines but mostly in lower-volume environments, and he thinks the technicians from the other teams do not understand what they are talking about. They lack the experience necessary to understand.

While Frank tells me this, he gets ready to apply a fix he developed to keep a baffle closed. Customers open it to clear paper jams and forget to close it; then the fuser cannot maintain its temperature and jams more. He uses a piece from a very old machine to make a spring to push the baffle back down whenever it is released. He puts a piece of heat-shrink tubing over the metal to spare users' fingers and holds the tubing in place until it shrinks with a red Ty-Wrap. The red ones are resistant to heat, unlike the white and black kinds. Frank would like to add a spring to the fuser to make sure it gets properly seated after being pulled out for jam clearance. An interlock would do as well. As it is now, the machine can be run when the fuser is not solidly in there, which will contribute to further jams.

At this point, Frank has done all the replacements and repairs this machine seems to need. The next step in the process is the xerographic setup, adjusting all the different components that contribute to making a good copy. In this machine the process is substantially automated; if all goes well, the technician only needs to enter some numbers. Frank knows the sequence by heart. While the process is running, he starts on the paperwork for the service call, listing the different things he has done and the parts he has replaced. Then the automatic xerographic setup stops; the display complains of Error 7. He looks up Error 7 in the documentation and learns that it means that the machine cannot adjust the copy contrast. The manual tells him to look at the xerographic system. Frank notices immediately that he had not put the cleaner assembly back in position, so he does so. He also checks that the instruments used in this process are set up correctly, although he thinks that the process would have complained sooner if they were not. While he is looking at that part of the manual, he reads some of the relevant diagnostic procedures to see if they suggest anything else he should check. They do not, so he uses the diagnostics to run the cleaner assembly to make sure that it is in a normal state, and then he tries the xerographic setup again. He says if the process is interrupted again he will change the batteries in his electrometer, one of the instruments used in this process.

The error recurs. Frank first checks the connection to the patch generator, one of the principal components involved in this process; he says they are notorious for intermittent problems. There are some flaky contacts inside; at one point the technicians were told to take the patch generators apart during service calls and solder those connections. Instead Sam, the team's technical specialist, started doing all of them on an exchange basis. Technicians could take patch generators Sam had fixed as long as they brought the replaced ones back to be fixed. Now there are new ones soldered at the factory. Frank decides the connection to the patch generator is good enough not to have been the problem, so we go out to the car to get a fresh battery for the electrometer. On our way back in, he notices someone copying engineering drawings and tries to interest him in a new machine that the corporation offers for such work. The response is that Frank is wasting his time; the power to make that decision lies elsewhere: "I don't have any clout; I just work."

Back at the machine, Frank changes his battery and tries the process again. He says if it fails again, he will give up on the automatic process and do it by hand, one step at a time. This would make it clearer what is failing and when. The machine takes a long time in the preliminary part of the procedure, before it starts moving paper, and Frank comments on it. There is another error that Frank is expecting to see, which will appear if the machine has not managed to set the contrast after a large number of copies, although the error message complains of something else. The machine runs out of paper before we reach that limit, and we joke about feeding it another forest as we reload.

Frank starts it up again, and the error recurs. This time he looks at the diagnostic procedures prescribed for this error code. The procedure tells him to check the values which the machine has reached in the process of trying to adjust the xerographics. He does. Because they are outside the acceptable range, the instructions tell him to set new values, different from the default values with which the machine starts the automatic process, and then to run the automatic process again. While it is running, he looks up the part number for the patch generator which he suspects. The one in the machine has a label on it indicating that it is one that Sam fixed, but they do not last forever. If this process finishes, he will leave a note to suspect the patch generator if there are further problems.

The process is still running. Frank shows me that the machine is displaying the values which it is setting, and he tells me that you can interfere in the process by telling it to accept the value at any time. As the machine finishes one part of the setup, he notices that the values are not the normal ones, but the book says they are in the acceptable range. We notice some odd copies; Frank thinks they happen because the machine is switching its settings from Darker to Lighter in the middle of a run, and these copies were made in midtransition. He tells me he will go back and check the Normal settings when it finishes. The automatic process does Normal first, then Darker and Lighter, and Frank thinks that the process of adjusting the latter two can change the first, which is the most important since most copying is done in Normal. He also likes to set the machine up a little lighter than is prescribed; he thinks it will run longer before needing another service. The xerographic setup finishes, and we are in business.

Frank starts setting the exposure, a matter of exchanging apertures and running test copies.[15] The light is good, and the machine sets up with a small aperture. The spare apertures are now hidden in a compartment on the door; when that change first occurred, no one told the technicians, and some of them did not find the apertures until after two or three calls on the new machines. Frank looks at the balance next. Bob has biased this machine to

---

[15] Exposure of the photoreceptor is controlled by the aperture, as on a camera. In this case, the apertures are long narrow pieces of metal with slits of different sizes running the length of them.

work better with eleven-inch paper; it would not work right with fourteen-inch, but Frank presumes Bob knows what his customers use. Finally, Frank goes back and sets everything a little bit lighter to make it run longer. He refers to this as a "Chicago Fix."

Now he is done; it is time to clean up and do the paperwork. The book says that replacing the retard rolls in the paper trays, the ones Frank replaced because of the oil on them, requires checking the paper tray settings. Frank figures that the couple hundred copies we ran in the setup process constitutes a pretty good check. He checks the billing meters. He checks the part of the machine log that shows how long various components have been in the machine and resets those he has changed. This also tells him how many copies he used during the service call, a number that is reported as a credit to the customer's account. He writes up the service log, noting particularly his adjustment of the setup values and the fact that he is out of territory. He copies the letter that technicians leave for the customer at the end of the call as a demonstration of the state of the machine; he notes that they need more fuser oil and more toner. Frank prefers that customers keep the supplies separate from the machine, so that he or someone he has coached will put them in, increasing the odds that the right supplies will be put in the right part of the machine. He has seen some strange combinations over the years.

Frank phones in to report the call; we have been there three and a half hours for a fairly minor problem and a lot of routine service. He is on hold, complaining about the music played for those on hold. He thinks most customers will not wait if they are put on hold; they do not like the music and feel patronized by the taped reminders that calls will be dealt with in turn. Finally he gets an operator. The reporting is primarily numbers; there are numbers for problems, numbers for the part of the machine where the problem occurred, and numbers for the parts that have been replaced. Then he checks for new calls; he only has one, the whole team only has five. It seems clear that my presence is not affecting the team's performance in any significant way.[16]

As we leave Frank says something about parts and LOLOS, so I

[16] The corporation worries about visitors interfering with service. If the statistics reported by a district with many visitors change significantly, the visits will be cut off.

ask about LOLOS. He cannot explain; it is the "blank blank Level of Service," but it is really the list of parts he is supposed to carry.[17] The form is made to serve as a guide to where the parts are stored in the van or car, as an inventory, and as a help in ordering new supplies. His district has customized LOLOS, allowing the technicians to set their own levels for parts they should keep on hand, and his is substantially changed from the original. The original was not very accurate, just an engineering SWAG. SWAG? "A Scientific Wild-Assed Guess."

Commentary

There are some interesting stories in this complex layering of narrative, including some stories about stories. Most of this vignette is my narration of the service call. However, one should note that when Frank talks to me, it is only occasionally about this specific call; his discourse is far more often about practice, his territory, or the history of particular features of this machine. The point is that this is a routine service call. Most of the work has been reduced to practice, and Frank need not concentrate on what he is doing to the exclusion of other thoughts. The call is unremarkable and not worth a story except where I need explanation. He constructs an account of this machine at the beginning of the call, using the logs and the narrative elicited from the users, and finds the machine to have a couple of specific problems, one from wear and one from misuse, and to need a great deal of routine service because of the time that has elapsed since the last service call.

The narrative of the users actually begins with some words from Frank about the difficulties of getting a useful narrative from users because of conflicting terminologies used for the machine. The part of his narrative about the oily retard rolls is constructed from the clues he finds in the machine and his knowledge of the common ways in which users fail to do their work correctly. The narratives that appear thereafter are either history or reflections on practice, until the xerographic setup process produces an error.

[17] LOLOS is the only acronym in the service world that I have never been able to have defined. This lack of absolute definition does not seem to bother the technicians at all, being one of many things in the world determined elsewhere.

Frank tries a series of procedures to dissolve this problem, assuming that the standard answers will take care of the problem without his determining what the problem actually is. These fail, but the documentation has a procedure that fixes the error. We are left with no account of this problem, and in a sense it is never actually diagnosed. Neither Frank nor I knew what the problem was, although presumably the designer of the documentation did.

Technicians cannot work only in their own territories; the pattern of failures is too irregular. When they call Work Support for assignments, they first check for calls in their territory. If there are none, then technicians accept calls for others in their subteam or in the team as a whole, in that order. Frank has no calls for his own territory, so he is taking a call for another member of his subteam. Access to the machines is a perennial problem for the technicians. In most of the corporations in Silicon Valley, employees either wear or carry badges, and access to the building is denied to others. However, there are people like the technicians studied here who need regular access to the building but are not employees. Two common strategies are to give them badges of their own and to maintain a number of vendor badges that are issued to persons showing identification from one of the appropriate companies.[18] Even with badges there can be problems: in the first vignette, Jim told Tom that one of the customer corporations believed that Tom had taken one of their vendor badges with him after his last visit. Tom replied that he never gets a badge there; although vendor badges are used at that site, this group of technicians is somehow exempt from the requirement.

This concern for badges is only a matter of normal industrial security. At other sites where defense work of various classifications is conducted, technicians may need to be escorted at all times by an employee, while red lights in the corridors warn other inhabitants that an uncleared stranger is in their midst. Unnecessary items, like ethnographers' tape recorders, may not be allowed in the building; technicians' tool bags are carefully inspected going in and out. Spare parts may be allowed in but not out. In extreme cases, as I mentioned earlier, the technicians are not allowed in at

[18] This does not apply to those who are observing the technicians; I always signed in and out as a visitor.

all; the machines are brought out. This can hinder the technicians' work, in that they are deprived of a chance to see the machine in context and may not meet the relevant users.

Once in the building, the technicians then need access to the machine and to the users. This may only be a matter of finding someone with the keys, as it was in the second vignette. Here, however, since this is not Frank's territory, he begins by finding the person responsible for the machine, in this case the account manager. When Frank took the call from Work Support, he got the machine's serial number, the address where it is located, and the name of the person responsible; sometimes this is the account manager and sometimes the key operator. Account managers either purchase or lease the machines and supervise the service contracts. Their power in the lives of the technicians stems from the fact that they can order the machines taken out at any time, but they may or may not actually use the machine. The key operators are a second important category of customers. Ideally they assist other users and generally track the state and use of the machine; the degree of their involvement largely determines whether the placement of the machine is a success or a failure (Blomberg 1987).

Account managers and key ops are also important to the technicians in that it is through them that the technicians have access to the users who have had problems with the machines. This access was important in the diagnosis of the problem in this case; the users' comments point Frank directly at the problem. It became clear during my fieldwork that this reversing-roll problem is well known and routine for any competent technician, although it is not addressed by the documentation. This was probably a very difficult diagnosis the first time, but it was virtually automatic for Frank here. The sort of confusion of terminology that Frank describes while we wait for the users is common in the interactions of technicians and users. It is best avoided by talking to the particular user at the machine, so both can look at it, point at things, and negotiate a mutual understanding, and this, in fact, happens easily in this case.

Frank reads the logs before and after the interview with the users but does not do so to confirm or deny their reports. He has already confirmed their report by feeling the free play in the shaft. His interest in the logs stems from his earlier observation that it has been a month and a half since this machine was last serviced. He expects this to be a major service; the logs will indicate which components need to be changed as a matter of routine and whether or not there are other problems to investigate that were not experienced by the users directly.

There are three logs involved in this machine. Two are maintained by the machine's software; the third is on paper and is kept in a compartment in one of the doors. In one of the software logs, the machine records routine events as a way of measuring the lifetime of components. In the second, it records various sorts of problematic events, some of which would have been perceived by users as jams or other troubles and others which the users could not have seen but which indicate that the machine's behavior is changing in undesirable ways. (There is another category of problematic event, those that are perceived by the users but not recognized by the machine, but these are only reported as service calls.) The log that records routine events must be erased deliberately, by a command from the control panel; this is done when the components involved are changed. The log that records problems is erased automatically when the technician leaves the diagnostic program. This is to ensure that the problems reported are recent events, so in theory, at least, the log only displays those events that have occurred since the last service call. In practice, there are ways to avoid erasing the error log if one does not want to take the time to write it all down on the form that is part of the paper service log. This might occur, for example, on a courtesy call, as in the third vignette, where one is not working on the machine but merely observing it. A technician would want to check the log on a courtesy call but would not want to erase it, so that at the next service call it would accurately portray the entire history since the last service call. The paper log is used to record the contents of the machine's software logs as well as things done to the machine, pieces replaced, adjustments made, and things to suspect the next time. This is known as the service log; the two logs in the machine's software are known to the technicians by the codes which bring them up. I will lump the latter two together as the machine logs. The logs, of course, do not necessarily reflect the users' perceptions of problems.

The issue of fending off would-be users is a constant one for technicians. The machine is at the customer's site for the customers' purposes, but they must give it up to the technicians for service. This service is scheduled without consulting the vast majority of users at a customer site, however, and some number of them will not be happy to find the machine unavailable on any given occasion. There is not much that can be done about this, but in turning them away the technicians do not wish the machine to seem too unavailable, since these impressions may remain when these same customers have to fill out the Customer Satisfaction Management Surveys which serve as part of the technicians' performance appraisals.

Technicians distinguish between casual and dedicated users of their machines. A casual user is anyone who uses the machine but whose job is not primarily running the machine; this presumes that the machine is available for such use. Dedicated users are usually operators in copy centers; their job is to run the machine. Technicians presume that dedicated operators will learn more about the machine, if only from experience, and may be expected to be more interested in learning. Casual users cannot even be expected to know how one is supposed to use the machine or even to know that there is something to know. Accordingly, Frank thinks of labels which might tell them what to do when they are there, or modifications which might force them to do the right thing. The spring he adds to the fuser is one such modification.

A persistent theme for technicians is the need to protect the machine from its users. Incorrect use of the machine can cause as many problems as component failures, but as Frank's tales of his own customers are designed to show, users may resist being taught how to do things that they think they already know. The technicians accordingly develop the ability to detect the history of the machine's use from the clues remaining. The oily retard roll suggests two different ways in which the customers might have been misusing the machine; either suggests to Frank an unwillingness to learn the proper way to operate the machine or the appropriate materials to use.

Problems stemming from users' misdoings are clearly identified by the technicians as products of human action, not as something gone wrong with the machine. A similar identification occurs with the switches whose mounting screws are too tight: someone tightened them too much. In fact, the presence of the machine in the customer's location is seen as the product of a long series of human decisions and actions in design, manufacturing, and sales, none of which are accessible to the technicians. The users are available, however, and the technicians try to manage the users' interactions with the machine through labels, springs that will not let the users leave baffles in the wrong position, classes to teach them the right ways to use the machines, or places to leave the supplies which are unavailable to the casual users. Frank admits that he finds the problem of misuse very frustrating. While he says he may overreact with customers in his own territory, he did not do so on this call, and his feelings of frustration were the same as those I heard about from other technicians.

Frank's comments about what users know and do not know, will and will not learn, are reflections on the social distribution of knowledge and how it is accomplished. His comments about team meetings address the same issue among the technicians. It is particularly striking that he feels the lack of comparable experience makes it difficult for the other teams to understand what his team has learned, so they cannot take advantage of his team's experience. Implicit in this are the technicians' assumptions that their skills are not learned in school but from each other, and that the meaning of their talk about their skills is not obvious outside the context in which they were developed. Frank was delighted that I had been to school on this copier.[19] He had been observed by an earlier visitor doing a time and motion study who refused to speak to him at all. Frank had wondered how the observer could write down what Frank was doing if he knew nothing about the job.

Frank began working with the corporation in Indiana, which may account for his reference to a Chicago Fix. Some districts get new machines before others, and some have accounts that will use

[19] This is the service idiom for having taken the course to learn how to repair this copier.

the machines much more intensely than others. These districts encounter most of the problems before others will and must develop their own ways to fix problems until engineering and manufacturing catch up. Word of these innovations will spread, usually carrying acknowledgment of their origin. Such districts will have a reputation among the surrounding districts as something like the big time, the cutting edge of the blunt instrument that is field service. Silicon Valley is such a district on the West Coast, Chicago is the same for the Midwest, and this may remind us that although service is intensely local, and it is the local geography that most concerns the technicians, they do this work as members of a multinational corporation, and what happens elsewhere in the corporate geography may matter to them too.

3

Territories: The Geography
of the Service Triangle

The social arena of service is defined spatially by the territorial divisions of the service world, and these territories are a fundamental category of concern for the technicians. Each technician has a territory, which has a given number of machines placed with a given number of customers. Each machine has a history, a life span, and a pattern of evolution as changes and modifications are required. There is a history, too, of the relationship with the customer, and neither technicians nor customers have permanent positions. Often a technician takes over a territory from a different technician, or acquires machines from several different technicians, inheriting at the same time the consequences of both the predecessors' work habits and their relationships with the customers. The technician is responsible for keeping the machines running, keeping them up-to-date, and performing maintenance at the appropriate intervals. The technician's responsibility to the customer is to keep them contented, which often involves more than just keeping their machines running.

This situation would be complex enough if the technicians were able to focus exclusively on their own assigned territories. However, a technician is also a member of a team which covers all or part of a district, and of a subteam which covers part of the team's area. Outside one's own territory, responsibilities are first to fellow subteam members and then to fellow team members. It is rare but not impossible to get farther afield than one's own district; at the local level it is routine for a technician to work on machines belonging to other members of the subteam or team. This means that relationships with the customers, as well as the state of the machines, will be affected by the performance of other technicians, while part of one's own work will be in others' territories, affecting their machines and their relationships with customers. It seems clear that technicians worry more about the social damage another technician can do in their territory than about what might happen to the machine, perhaps because the machine would be easier to repair than the delicate social equilibrium.

At the same time, technicians do not feel responsible for doing as much work on the machines during service calls in others' territories. Their theory is that they will repair the immediate problem, while retrofits and some long-term maintenance tasks are left for the technician to whom the machine is assigned. Nor do they seem to worry about the social equilibrium of another's account; indeed, they cannot. Although everyone works out of territory, it still happens that the technician responsible for a machine sees it most consistently. This technician provides the continuity and bears most of the burden of the social relationship. In return, it is this technician's performance appraisal that is affected by the customer. Normally, other technicians visiting a machine will not have been there often enough to be a significant part of the social relationship.

The social dimension of service is not well recognized. The technicians know it exists, and their immediate management seems to know this also. There is little sign that the customers recognize it. Technicians' work is commonly defined as the diagnosis, repair, and maintenance of machines, and technicians servicing others' machines take advantage of this. They go on such calls to do work under that definition, playing out the script that service is just about fixing machines, being polite to the customers but keeping their distance.

One could seek to protect relations with one's customers by attempting to prevent others from working in one's territory. I have heard that some technicians have told specific individuals to stay out of their territory, but one cannot keep all others out. Nor can one work only in one's own territory. The workload is too variable, and technicians will all have times when there are no calls in their territory, and so they must work elsewhere, and other times when there are too many to answer, and so they must accept help.

A different option is for a subteam to regard their various territories as essentially one and meet frequently to share information to keep each other up-to-date. There is some inherent tension between this collective approach to the work and the assignment of individual responsibility for every machine; each technician must either believe that the other members of the group are as skilled and thorough or be willing to accept the possibility of degraded performance in return for greater resources. The subteam described in the first vignette shares territories in this way, with the added incentive that they are also collectively responsible for a small population of older machines that no one else in the team could fix. Consequently, it behooves them to take care of each other because no one else is going to help with those machines. Technicians without such constraints have a wider pool of potential help and so can be more casual and more selective about alliances and less bound by the subteam structure.

The contradictions endemic to the service world also appear in the technicians' sometimes conflicting reactions to their territories. Territorial boundaries provide a reference for all action, whether one is inside one's territory or outside. There is a desire to make these boundaries rational and to live within them and a recognition that this will not happen. The technicians express a desire for geographically compact territories to minimize travel time but do not want to surrender their more far-flung customers until they can find a technician fit for both the technical and social challenges. A technician may not want to be responsible for a customer's new machine in a distant location outside the technician's territory, but customers often want a single technician to service all their machines. In such circumstances, the technician cannot resist too much without risking damage to the relationship with the customer, and so may instead appeal to the team manager to deal with the situation. The team manager divides the district into territories and assigns them to the technicians but then needs to adjust for new machines and new technicians or for the loss of either. In this process, machines may cease to be assigned to any particular technician for a period of time; this usually results in progressive deterioration, as immediate problems are fixed but no one is responsible for the machine's overall well-being. When this oversight is caught and the machine is reassigned to a technician, the effort required to restore it to good condition may be formidable.

For all interactions in the service world, one factor is the issue of territory—whether or not a machine is assigned to the technician in question and where it is geographically with respect to her (other) assigned machines. Whether a machine is in-territory or out-of-territory changes one's responsibilities to the machine. It also changes the performance expected by one's teammates, and it crucially affects one's relations with the customer.

4

The Technicians

The real work of field service technicians is to maintain a triangular relationship between the technicians, their customers, and their machines, and each group of participants will be discussed in turn. However, everything here is told from the perspective of the technicians, and it will be the technicians' views on customers, machines, and relations between the two that will be reported. It is the technicians, too, who find control to be difficult and ephemeral in this domain, something for them to reestablish in the specific situation, knowing it will only be lost in the use of the machine, and it is their understanding of the machines that is fragile and contingent, known to be true in specific circumstances but uncertain in further application. Moreover, it is their stories about their actions in this world that provide the substance of this work, so it seems appropriate to begin our discussion of the population of the service triangle with the technicians.

There are two striking characteristics about the background of technicians as a group. The first is that a surprising number come from rural backgrounds; one-fifth of my copier class had grown up in homes without electricity or indoor plumbing. The second is a propensity to tinker, and this was as true for the women as for the men. Working on cars in one way or another was a common theme. Some reported helping their parents tune the family car; another helped in his father's junkyard and at age thirteen sold bicycles built from junk parts. This characteristic tinkering continues: the technical specialist of the team I observed most is famous for taking parts home and rebuilding them.

There was nothing particularly distinctive about the educational backgrounds of the technicians I observed. All had high school diplomas. Approximately half of the technicians in the district at that time had junior college degrees in technical subjects. About twenty percent had learned their technical skills in the military, a fraction that was greater in the time of the draft and the Vietnam War. Most of the remainder had acquired their technical education (other than that provided by the corporation) from private technical training institutions. Some four or five percent of the technicians had bachelor's degrees, but these were all in nontechnical disciplines such as English, psychology, or drama.

Technicians work in a district and are divided into teams, primarily on the basis of machine type, so that a team services machines of similar capacity or technology. Teams are divided into CSTs, or subteams, on the basis of shared scarce skills, or more commonly, territorial proximity. Membership categories in the teams are manager, specialist, senior technician, and technician. The managers are usually promoted from the ranks of the technicians, but few succeed in the job, and most return to being technicians. The specialists are also promoted from among the technicians; their job is to monitor and improve technical skills, assist with difficult problems, and circulate technical information. The division between technician and senior technician is just one of seniority and salary; reputation and skill are not thought to be related to this distinction.

There are few career paths within the corporation for technicians. Despite the low success rate, many do try to become managers, although the technicians say that doing so is a waste of time and entails losing touch with the real world of service, trading a real job for headaches. There are fewer opportunities to become a specialist, and most of those who get the job stay there. Specialists are still in the community of technicians. However, some specialists get promoted to positions as trainers or as regional or even national specialists. These are definitely outside the community, and there are very few such jobs. There are no other career paths within the corporation built on work as a technician; one could imagine abandoning that experience and starting over in some completely different capacity, but it is not clear that this has ever happened.

The organizational arrangements of the team structure within the district do affect the work of service. The initial loyalty is to the subteam; the question "Are we clear?" (i.e., "Are there any calls for us in the queue at Work Support?") refers first to the calls for the subteam and only secondarily to those for the team as a whole. Similarly, at least some of the technicians feel that the subteam is the most appropriate place to look for help and that it is not appropriate to expect help from other subteams.[1] Team and subteam organization are not constant, however, and new machines and new or retrained technicians mean the social situation is continually in flux. While technicians worry about the assignment of new machines, they do not seem concerned about the actual installation. Most "installs" at the time were done on overtime by whichever technicians volunteered for the extra work and pay. This was done so that technicians could address service calls during the day and not be committed to a long but normally straightforward procedure. Doing installations this way also serves to minimize the impact on the customers, since the work is done after hours.

Teams do show some sense of unity, and the better technicians are free to float through most of the territories without regard to subteam membership. Most technicians enjoy informal, casual relationships with their immediate managers, most of whom have recently been technicians themselves. Some technicians did wonder if their manager would be interested in my tapes, but immediately turned the comment into a joke about the manager's height: "All he'll hear are short jokes." Movement of technicians is not just within the team. Teams servicing more complex machines recruit from those servicing the simpler ones, and teams that are overstrength for their machine population may lose technicians with appropriate training to teams with too few technicians. The team is the smallest formal organizational unit; the subteams are

[1] Actually, it is not at all clear that these technicians did not themselves ask for help from outside their subteam; it is only clear that they regarded some requests for help from outside their subteam as inappropriate.

too ephemeral. Consequently, the team is also the real meeting point for the technicians and the corporation. The team manager must not only represent management to the technicians but must also represent the technicians and the realities of the service world to the corporation.

In that world, the technicians respond to their problematic service situation in a collective manner, tracking each other, sharing information about particular machines and about machines in general, and keeping track of events in the world around them which may affect their work. The organization of the technicians into teams and subteams affects but does not determine the collective response; the official structure is invoked or ignored by the technicians as seems appropriate to them. A distinctive aspect of collaboration in service involves the technicians' appraisal of other technicians. They need a reasonable understanding of each other's strengths and weaknesses for two principal reasons. The first is that another technician may solve a hard new problem; one must know something about how that technician works in order to know what to make of the account one hears of the solution. The second reason is that other technicians will take calls in one's own territory and one needs to be able to anticipate the effects, both on the machine and on the social situation.

No technician works completely alone, but the amount of co-operative work varies over a wide range. At the opposite end from those who try to work alone and only in their own territories are members of groups that effectively share territories. These groups may align with the subteam structure or may be a subteam with the addition of friends and allies. Persistence of these alliances through time varies; the group in this study which seemed the best established also shared responsibility for a small population of different machines. Only members of the subteam could help each other with those machines. This incentive seemed to encourage their group collaboration, but they did still cooperate with others. They shared more information than the less intense groups, in particular watching the coverage of their common territories. The more casual groups tend to focus on problems or unusual occurrences, and in fact, groups may coalesce around an unusual problem. A technician arriving at a popular lunch spot with a problem will quickly engage the attention of most of those technicians present, as in the case presented in the fourth vignette.

Technicians try to know what happens to the machines in their territories, such as what failures have occurred, who has worked on them, and what modifications have been done to them. If the opportunity occurs, a technician who has taken a call on another's machine will report to the responsible technician what was done and what was observed, particularly anything which appears likely to become a problem. This reporting is most detailed between technicians not in the same group, where the technician who took the call is unlikely to work on the machine next and is also not expected to do much more than fix the cause of the service call. Within a cooperative group, each technician is expected to do whatever is necessary for the long-term health of the machine, including maintenance, retrofits, and repairs in anticipation of impending trouble. Consequently, such a group will talk more about who was where, implicitly assuming that everything necessary was done. This assumption holds for group discourse even when the members have reason to doubt its validity.

The conversation of a closely cooperating group can be quite cryptic when the members are sharing information about work; they are considering a well-defined field which can be discussed with considerable economy, verging on code. The work is discussed in terms of calls, the units of work that appear on the computer screens in the Work Support Center and the units in which it is done and reported. "Call" refers to both the call for service from a customer and the service call or visit to be done by the technician; these calls are the events or episodes defining work in the service triangle, even though problems may persist over several different calls. In technicians' conversation, calls are sorted according to the technician primarily responsible for the machine and are discussed in that way, even if some other technician takes one of them. It will still be a call belonging to the assigned technician, as in "I took that call of Jane's." The machines are otherwise identified by the name of the corporation where they are located, a street name if the corporation has more than one site, and possibly a model name if other than the most common machine. For a technician who knows the territory, this terse identification includes the significant history of the machine and details of the social setting, including the dominant personalities. This can extend to machines that are not yet installed but are going to a company that already has machines; the personality of the customer's manager in charge of machines will be expected to influence the servicing of the new machine as it has the old.

Discourse, then, within such a group will emphasize the problematic elements. A trusted colleague taking a call on a machine that works well at a site that works well gets minimal comment: "She took my Fairview city hall call." Minor discussions may focus on clarifying who was where or reconciling different understandings of a third party's activities, wondering at the absence of a regular member of the group or expressing frustration at failures of communication, both external (such as trying to reach a colleague through a customer's phone system) and internal (for example, being assigned calls to customers who were observing a holiday). These are routine matters, handled with dispatch and leavened with a bit of teasing and mild character assassination. Efficiency in dealing with the movements of people among places permits the technicians to focus on their real interest, problems with machines.

Technicians will ask their colleagues about a specific machine that they know to be troublesome if they have observed a call up for that machine.[2] Recurrent problems are intriguing both for the failure of the technicians' collective knowledge to date, shown by the growing list of fixes that have not solved the problem, and for the tension in the social arena as the customer's irritation grows. Discussion of a good baffling problem will involve all the technicians present, whether members of the responsible group or not. There is an assumption on the part of all participants in the service world that the technicians can solve any machine problem. The failure of this assumption in a problem which defies repeated attempts at solution is thus a serious challenge to the competence of the community and so engages their attention.

In the fringe areas of their collective knowledge where the technicians do not have enough information to identify causes, problems appear as symptoms susceptible of multiple interpretations

[2] That is, listed in the Work Support queue.

with no good way to discriminate among them. The technicians have an incomplete connection between the multiple interpretations and the set of fixes known to cure some manifestations of the symptom. A solution to a difficult problem could either increase the number of interpretations for a given symptom or clarify the connections between that symptom and its manifold causes. Either is acceptable; the alternative, an insoluble problem, is not.[3] Until a persistent and difficult problem is solved, it will be a challenge and a source of worry to the community of technicians, mingled with the tantalizing and appealing prospect of learning something new about the machines.

An intense group discussion of a machine and its problem occurred over lunch one day. A technician and I had just come from a machine with a recurrent, particularly opaque problem. There had been several service calls on it already, and most of the known fixes had been tried. The machine in its failed state appeared to provide no useful information about the problem. Two of the other technicians from the same subteam also showed up for lunch. A particularly skilled technician not formally part of their group was working with one of them that day and so was available to provide a fresh perspective. Introduction of new persons contributes to the collective consideration in that the summing-up necessary for the new participant to understand the situation permits those who have been working on it to look again at the situation as a whole. Since the difficulty in solving such problems lies in the weakness of the diagnostic information, the summing-up consists primarily of listing their attempts with different strategies known to dissolve the particular symptom. Such a summation is not entirely satisfactory because some of these dissolution strategies are parts replacements, and the technicians know well that even new parts are not necessarily reliable.

The group's attention was turned to this machine by an inquiry from one of the other technicians, who had noticed a call up on it when talking to the Work Support Center. In the discussion that followed, the new participant initially was in the position of sug-

[3] Quite literally not acceptable. If a machine with a maintenance contract cannot be fixed, it must be replaced at the corporation's expense. This like-for-like exchange, as it is called, can also be expensive in terms of professional prestige.

gesting the best-known causes for the machine's behavior, all of which had already been tried. This dialogue helped to establish that those working on the machine had behaved as competent practitioners, in that they had pursued all the known solutions to this problem. The fact that they had repeated some of them a couple of times does not detract from this competence, because replacement parts are known to be faulty at times. Part of the discussion addressed which, if any, of the error reports from the machine could be considered relevant. Competent practice includes discarding diagnostic information from the machine when it tells nothing useful, and hence part of being a competent practitioner is being able to distinguish those times from the others.

In this particular case, all parties to the discussion had a common understanding of the possible causes for this problem, and those working on the machine had addressed most of them or were planning to do so. The new participant suggested a novel diagnostic strategy, which most of them had heard of before and discounted as being incredible. It did, however, address the problem as they understood it and promised a more subtle discrimination among its multiple causes. The strategy involved the use of a very cheap FM radio to detect electrical noise, making a virtue of the cheap radio's inferior filters. Although the discussion at lunch appeared to conclude with a reluctant acceptance of this diagnostic technique, what had actually been accomplished was to confirm their common understanding of the nature of the problem. Subsequent efforts to fix the machine addressed that understanding without the radio being used.

By contrast, certain parts are known to produce the symptom at issue, although there was no diagnostic evidence to connect them to this manifestation. One of these parts had been replaced twice in recent months, and the technicians were reluctant to acknowledge the argument from experience made by one of their number that this part should still be suspect. However, at the end of the lunch and the discussion, one of the technicians offered another of that part to the technician working on the machine, who hesitated, expressed hope that it was not the problem, and took it.

In effect, then, the members of the subteam reviewed the history of their actions with the machine and their current understanding of the machine and found the two consistent. Given what they knew of the machine, there were still strategies to try, both dissolution through replacing parts and diagnosis with the radio. They questioned the adequacy of their understanding but had no information with which to modify it; the group had tried to create a better account of this machine but still had no new information which would improve the coherence of the narrative. One of the requirements of field service is that calls must be taken. If the machine is not working and the customer has noticed, something must be done to the machine. Under these circumstances, the members of the group consult each other, put together their best understanding, and set out to do what they know how to do.

In addition to sharing territories and pooling information, groups that work together share parts. It is very difficult for a technician to have all of the necessary parts all the time, and the Parts Drop does not always have everything in stock. While the team technical specialist does keep a stock of particularly problematic parts, to some extent technicians rely on their colleagues to fill the gaps in their own collections. This is done selectively; in general, the people one borrows parts from are the same as those one asks about difficult problems.

Groups that work together also share scheduling, teasing, and a tolerance for each other's weaknesses. They plan collectively to deal with the calls facing each of them. They tease each other about their probable fates, professional and personal. They will admit, very privately, that some members of the group do not work as hard as others and have not necessarily done what they say they have. There is no public reproach. As with so many other factors in service work, the other members of the group can do little about such a situation, so the objective is to find the least painful way to work around it.

In general, technicians' attitudes toward other technicians combine honest appraisal with acceptance. Technicians can neither choose nor change their colleagues but will nonetheless be affected by their colleagues' work habits and social graces. Technical weakness is better accepted than laziness, which in turn may be mitigated by personal charm. In general and within certain limits, members of a subteam tolerate each other's sins better than they do those of more distant colleagues. Conversely, technicians seem to express negative opinions about distant colleagues more freely than they do criticism about those with whom they must work on a daily basis.

For example, the revelation that a teammate had not done a needed retrofit was greeted with surprise, but not with censure. On another occasion, it was discovered that parts reported to have been replaced did not appear even to have been cleaned, but this discovery was not regarded as worth mentioning to the technician responsible. However, other technicians in the subteam displayed high anxiety at the possibility that the technicians responsible for these omissions might visit customers with whom relations were more delicately balanced. This relative tolerance within was balanced with a sterner face without. The same technicians reacted negatively to a request for help from a technician not part of their subteam, commenting at great length on the unprofessional behavior they had observed of that technician and dismissing the situation as no crisis but merely a normal backlog. Similarly, another technician not of their group was said to get overexcited and to abandon the systematic approach they believe characterizes competent practice. However, they did seem to think there was hope that this technician might learn professional behavior. It is unclear whether this hopeful attitude was occasioned by any actual difference in expertise or practice, or perhaps more by the fact that this technician's territory adjoined theirs while that of the more scorned technician did not. Proximity involves more interaction, more floating in and out of each other's territory and hence more actual impact on each other's working life. In my observation, both offending technicians seemed to be no less professional than any other technician, nor did their immediate colleagues seem perturbed by the behavior so offensive at a distance.

If the nature or identity of bad technicians is difficult to verify, because they always seem to be someone and somewhere else, there is little doubt about the good technicians. These are the technicians to whom others turn when stuck, who are expected to know what to do and to be good at it. The technical specialist on the team I observed most had a particularly good reputation for being able to find needed information about the machines. In the course of training and business visits, this particular specialist had formed a network of connections throughout the corporation which allowed him access to the information he wanted. This only added to his technical glory, already bolstered by a reputation for speed and a great repertoire of known problems and fixes. Other good technicians do informal consulting; a technician's choice to ask another technician as opposed to the team specialist may be based on proximity, availability, or comfort of personal relations.

There is a certain competitiveness among some technicians. Sometimes this appears in claims to the ownership of certain fixes, a claim that may backfire if the technician to whom one is speaking does not have a good opinion of the fix. At other times, the telling of a problem or fix is a challenge: Do you understand what I'm telling you? Have you heard of this before? Do you recognize the symptoms? Sometimes the telling of a procedure or fix takes on aspects of a verbal duel, in which the technicians take turns telling parts of a procedure, competing to see who can remember the smallest details or the many ways it can go awry. Technicians whom one would not expect to drop things will seem to brag about the number of times one can drop a part or an assembly in order to emphasize the difficulty and awkwardness of the procedure under discussion. One final appeal in such a duel is to claim to have just performed the procedure at a specific time and place, a claim that could presumably be checked, although doing so is probably unthinkable. Finally, even a good technician may go too far and be regarded as overzealous. I heard stories of a technician who was said to have addressed problems of loose fit with such ferocity that the next technician in the territory was unable to disassemble the repaired pieces.

The technicians should be viewed as an occupational community (van Maanen and Barley 1984). They are focused on the work, not the organization, and the only valued status is that of full member of the community, that is, being considered a competent technician. In pursuit of this goal, they share information, assist in each other's diagnoses, and compete in terms of their relative expertise. Promotion out of the community is thought not to be worthwhile. The occupational community shares few cultural values with the corporation; technicians from all over the country are much more alike than a technician and a salesperson from the same district. The technicians, however, depend on both home and client organizations for their own identities, one to provide the machines and pay their wages, the others to provide an arena wherein they may practice. The only real career option, promotion to management, means leaving the community, and most technicians would rather remain a technician-hero than become an organization manager. Then too, the technicians are in some ways more involved with their customers than with their own corporation, so even though they are always working in space that is not theirs, it makes sense to remain within the triangular relationship of technicians, machines, and customers.

5

The Customers

The technicians have a generally good relationship with their customers, those who use their machines. This is a little surprising in that most of the interaction between technicians and customers occurs when there is something wrong with the machine, which means a disruption to the customers' work. Technicians are outsiders in the offices where they work, and the presence of an outsider is something of a disruption, although it may also constitute a diversion, someone new to talk to. Moreover, the technicians represent the corporation responsible for the broken machine that is keeping work from being done, but they are also the best hope of fixing it and getting the work moving again. From the technicians' point of view, the customer who calls may even have broken the machine; it is fairly certain that they do not share the technicians' concern for and fascination with the machines. On the other hand, by having the machine and calling for help, the customers offer the technicians the opportunity to be technicians.

The basis of this relationship seems to be something like a social contract about the machines the customers have brought in for their own purposes. The customers then agree to admit the technicians, to give them the necessary help, and to tolerate the disruption of an outsider who may need information or other forms of cooperation. The technicians in return promise that they can and will solve the customers' problems with the machines and, at the same time, will disrupt the customers' business as little as possible. The customers do this to get working machines; the technicians gain the opportunity and place necessary for their work and thus, in some sense, their jobs.

The technicians, however, know that their work is not just the repair of broken machines and have sayings like "Don't fix the machine; fix the customer!" Within the triangular relationship of service, technicians focus on maintaining the relations between their customers and the machines, and this is accomplished through the technicians' relationships with both. One of the technicians' basic premises is that the machines will fail and need repairs; another is that the customers' understandings of the machines will differ from those of the technicians. A third is that the technicians themselves will fail, in that some of their attempted repairs will not work, although they believe that ultimately they will fix anything. To maintain some degree of control in the midst of endemic failure, the technicians feel they need to project a credible image of some authority, and they think of this as "being professional," in a strictly colloquial sense. There are two principal components of this image of professionalism. The first is that the technicians feel they should be recognized for the technical skills, understanding, and education necessary to their work;[1] the second is a desire to be seen as being businesslike, which seems to mean having it understood that one is concentrating on work in a narrow sense when making service calls. Carried to extremes, this emphasis could create some tensions in the social relations essential to that work, but in practice the technicians seem to try to maintain an appropriate balance, being both properly social and properly focused on the job, and the balance point shifts with the situation.

The technicians have little to do with initiating their relationship with their customers; that relationship is set up by the salesperson who places the machine. That is, the salesperson tries to sell the customer a machine that will do the anticipated work without exceeding the available budget. Given the reality of bud-

---

[1] This creates some interesting tensions. A desire to be seen as skilled involves getting or creating recognition of those skills from people outside the community, who may or may not want to admit that the technicians do anything special.

gets and planning, fairly often this results in the acquisition of machines that are somewhat marginal for the real workload. It also happens that some machines are simply easier to use and thus may draw work away from more powerful but perhaps more awkward ones. The service technicians tend to blame the sales force for not preventing these mismatches, and it is considered something of an achievement for a service manager to persuade salespeople to come to the service team's meetings and hear their complaints. The goal of these meetings is to avoid unhappy triangles of customer, technician, and machine; however, given the pressures that create these situations, the technicians are not optimistic that such meetings will accomplish this.

The customers own or lease the machines and conduct business in the premises where the machines are placed. They also grade the performance of the technicians through surveys distributed by the technicians' management. The corporate criterion for success in the service world is partially expressed in terms of the machine and partially in terms of customer satisfaction. Accordingly, the customers' attitude toward and understanding of the machine and their relations with the various technicians servicing it become matters of great importance to the technician responsible for the machine.

Customer satisfaction is measured through Customer Satisfaction Management Surveys. Periodically these are mailed to the individual at the customer site who is officially responsible for the machine. That person may fill out the survey or may pass it on to the key operator or to some other individual believed to have the best perspective on the machine. The relationship between the individual filling out the survey and the machine is recorded in the survey. If the review is bad, the process is repeated within six months. The technicians complain that these surveys often end up being filled out by the wrong people and that there is no assurance that an accurate picture of the situation is presented.

From the technicians' perspective, the wrong people are those who do not discuss the machines in the same terms the technicians use and who therefore neither understand nor describe the machine correctly and will not understand what the technician says about the machine. Not knowing the technicians' language for the machines indicates that these wrong people have had little contact with the technician, since part of the technicians' mission is teaching the customer how to talk about machines. Each technician has a favorite horrific instance of the ways the survey process can go wrong. In one case, all the copy center personnel at a local law firm were fired and replaced; the new supervisor then received a CSMS form and filled it out, giving the technician terrible reviews. The technician's principal complaint was that the supervisor did not even know him. From the technician's perspective, there was no longer anyone at the customer site who should have been filling out such a survey, since they had no experience to make them either satisfied or dissatisfied, but this is not a viewpoint shared by the corporation.

The corporate perspective is that there is no wrong person: "In measuring customer satisfaction, there cannot be a 'correct' or incorrect person to respond to the survey. Rather, we must consider *any* customer feedback as valid information" (*News & Digest* 1988; emphasis added). *News & Digest* is a journal published by the national service organization of the corporation for its members. This quote is from the response to a letter from a technician complaining about the survey system; the response was written by a manager in the service organization. This response effectively denies that there is anything to know or understand about the machine in order to make valid criticism and discounts a major source of trouble for the technicians, the fact that the customer must be initiated into the technicians' community of discourse in order to communicate about the machine.

It is unclear why managers privilege the survey results in this way. Managers primarily know about technicians' work through the technicians; they are too busy to spend enough time in the field to observe it directly. It may be that emphasizing the satisfaction of the customer over the performance of the machine is a way to establish some control in the service situation, and indeed, truly measuring the performance of the machine would be difficult. The nature of the work and of the requisite knowledge make control difficult; management can neither abstract and control the knowledge nor direct the worker. However, customer satisfaction surveys are a source of information about the work which is *not* mediated by the technicians; by making them preeminent, ignoring the protests of technicians when the surveys are filled out by the wrong people and ignoring the issue of whether there is real information to be gained from such surveys, management gains a measure of control, or at least a counter to the technicians' mediation of all other information about the work. Only a measure, however, because the customers' satisfaction still requires a working machine, more often than not, and control of the technical knowledge required to keep the machines working remains out of management's reach.

From the service managers' point of view, the surveys can be seen as a response to a dilemma. They need to know and show to higher-level managers whether work is being done in an area where they cannot control the worker's knowledge or the worker's schedule, and where there is no particular measurable output. Under the circumstances, knowing which customers are happy and which are about to cancel must seem like a reasonable concern; from the corporate perspective, continuity in the contract with the customers is probably the only significant sign of success in service. Whether the surveys actually convey any real information about the attitudes of the relevant customers is another issue; managers seem to believe that they do.

The need to initiate customers into the technicians' community of discourse contributes to a situation termed "customer perception problems" by technicians. (One cannot imagine a customer using this term.) It is a point of faith for the technicians that when the customer's view of the machine differs from theirs, the customer is wrong and needs to be taught how to view the machine. However, the responsibility for doing so rests with the technician and is part of what is referred to as "fixing the customer." In its narrowest form, "customer perception problems" would mean that the customer perceives something that is not so, such as that the machine is not functioning normally when it is. In a broader sense, it refers to all the ways in which the customer's understanding of the machine differs from the technician's.

The machine is embedded in the social environment and work activities of the user site. The customers tend to discuss the machine in language reflecting their perceptions of its use and its various functions; without training, in the technicians' view, they will not know the proper names of its components or functions or the technical descriptions of its failures. Given that the state of the machine is often not obvious but must be interpreted, the technicians need more from the users than the discourse of use provides. Therefore, the technicians need to teach the users both what to notice about the machine and how to describe it, so they will collect the right information and present it in a form useful and meaningful to the technician. This can only succeed partially, since the user will not become a technician and share the technician's experience and values. Under the circumstances, mismatched perceptions, influenced by different ways of understanding the machine, are probably inevitable.

The language issue affects the process of service calls in that technicians do not accept the customer's description of the problem without considering whether the customer knows how to talk about the machine. If the problem is not obvious to the technician, the customer reporting the problem has to be found, and a dialogue ensues to determine what was meant by the report. The differences in understanding also affect the language used with customers. Unless the customer is known to be competent both to notice the right things and to express them in the technicians' language, technicians will avoid technical terms and repeatedly check for comprehension to make sure that customer and technician are talking about the same thing.

Adjusting customer perceptions is a social skill cultivated by the technicians. A major component is projecting the image of a competent practitioner, a professional in their terms, whose credibility on technical matters is unquestionable. This image requires that one look like one knows what one is doing at all times, perhaps most critically when one does not. If, for example, a technician should inadvertently go to the wrong machine at a customer site, the technician cannot simply walk away and go to the correct machine but must do some minor service before leaving. Technicians try to make short service calls and to avoid "broken calls," for which they must leave and return. In both cases they wish to minimize their presence and the awareness of their presence at the customer site; the broken calls also look unprofessional because they suggest that the technician either did not have what was necessary to fix the machine or was unable to do so in the time available. Tidiness is another technique used both to minimize awareness of presence and to appear in control. Technicians try to keep their tools and the pieces of the machine in order, not strewn all over the area, and generally try not to create messes in the course of what is potentially very dirty work. All these efforts to appear a skilled and competent professional reflect the technicians' drive to be professional; the appearance may even help them retain control in those situations where it is most marginal.

This desire for a professional appearance is one reason that using a machine beyond its designed capacity distresses the technicians. Not only will the increase in routine service and machine failures create more work on the machine, but the increase in service calls will detract from the technicians' image of competence. Concern for their image is also part of the technicians' frustration with the sales force. If the salespeople do not properly match machines with customers, the service force will be in the position of trying to make the machine be what the customer wants, rather than what it is, and the greater the mismatch, the greater the effort that will be required. Anything that requires technicians to spend more time working at the customer site reflects on their apparent skill. It suggests that they cannot control the behavior of the machines, and this doubt may diminish their ability to control the customers.

The ability to manage both machines and customers is an important part of one's reputation among other technicians. One technician, in conversation with other technicians, reported that a customer had been asking him to take responsibility for a machine the technician thought was out of his territory. He claimed to have resolved the situation by "acting stupid"; my tape recordings indicate that both parties dropped the subject without any obvious stupidity on either side. The result is the same, in that the issue is unresolved, but in the told version, the technician determined the customer's perception of his behavior, and in the recorded version the customer's perception is not known but clearly not controlled. The technician's construction of self requires being in control of the situation. It also appears true that acting stupid, in the sense of claiming not to have the necessary information or authority, is perceived as a viable option to avoid the responsibilities of the situation or to avoid unwanted confrontations with the customer. In fact, almost any strategy is acceptable if the situation is sufficiently desperate. Another technician, working with a customer with a reputation for being difficult, took great pains to establish a relationship of common humanity, conversing on family issues when possible. The point here was to claim an identity beyond that of technician, so that if the image of professionalism disintegrated in an uncontrollable situation, the technician could still be seen as a parent, and therefore similar, rather than merely as a deficient vendor of services.

Residents of the customer domain are not all equal. The principal person from the technicians' point of view is the account manager, also known as the decision maker, the person who decides whether the machine stays or not. Most higher-volume machines are leased rather than purchased, and the lease may be summarily terminated. Decision makers may not actually use the machine very often, so their information about its performance is apt to be secondhand. The actual users comprise a widely varied class with a comparable spread in status. Dedicated operators are hired to run the copiers; key operators have other jobs that include some responsibility for the copier, such as jam clearance, problem reporting, and some assistance to other users. The casual user is anyone who walks up to the machine to use it. Decision makers are usually managers, of somewhat higher status than technicians. Dedicated operators and key operators, usually clerical personnel, are perceived to be somewhat lower status than the technicians, at least from the technicians' perspective. Casual users may be anyone from the lowest employee to the company president, a range to which the technicians must be somewhat sensitive.

As one might expect, relationships with decision makers are characterized by somewhat more tension than those with other customers, even those of high status. The technicians also feel a certain ambivalence because they must win the approval of the decision makers while believing that they do not really understand the machine. Decision makers are placed both to make demands on the technician and to be heard by the technician's manager, and so technicians and their managers pay attention to these people. They take them out to lunch; they stop by and ask how their machines are doing; they are extremely careful to keep them informed of developments with a difficult machine which may not be fixed and which has required or may require repeated visits for a given problem. This attention is increased if the account has several machines, to the point that there exists a special section of the corporation to take care of the needs of customers defined as major accounts. Major accounts may also get a resident technician whose principal assignment is to service their machines.

Accounts possessing multiple machines but not enjoying major account status may still prefer to deal with only one technician. This can pose a problem for the designated technician, in that this preference may conflict with the common desire for a compact territory. In the case mentioned earlier, the technician already believed that the customer was geographically far afield with respect to the technician's other accounts. The dilemma was compounded because the customer's desire to have the technician service a new, even more distant machine was expressed during a service call on a machine with a continuing problem that had not been fixed during several service calls by various technicians. The technician had to try to project an image of professional competence in the face of a situation challenging that image, while fending off responsibility for the new machine without offending the customer. The solution was to address the problem machine, dropping the discussion of the new machine. The technicians see this choice as the most professional because their values assign the highest priority to fixing the problem. The issue of machine assignment could be dealt with better backstage, away from the customer, by arguing for territorial integrity with the manager of service, who would in turn deal with the customer.

There is also an ambivalence in the relationship between the technicians and the operators and other lower-status customers. From the customer's point of view, the presence of the technician disrupts the orderly flow of business, but it is the only way to get the essential machine fixed. The disruption may also be a welcome break in routine for the operators, unless it creates too much of a backlog in work. From the technician's perspective, users are apt to abuse or break the machine, but they have the information necessary to fix it. Interaction is more frequent and easier with these actual users than with the higher-status decision makers. Such users are also likely to be more familiar with the machine and thus more likely to know the language to use in discussing it.

Dedicated operators, key operators, and casual users of lower rank are also closer in status to the technicians than are the decision makers and casual users of higher rank, although the exact relative standing is not clear. They often share the status of not having their skills highly regarded by their management. Technicians claim to be skilled, but the claim is problematic in that service management believes the technicians can be replaced by semiskilled labor with a good set of instructions. Clerical workers claim to be more skilled and responsible than their management appreciates. If the technicians are perhaps more obviously skilled, they also get dirty in the exercise of that skill. If clericals are less skilled, they stay clean. All these issues serve to confuse their relative status and thus their relationship. In this situation of similar but ambiguous status, it is not surprising that users and technicians have developed a teasing style of dealing with all the critical issues of their working relationship, a style remarkably similar to the classic joking relationship (Radcliffe-Brown 1950; Evans-Pritchard 1951). They tease each other about the ways that users do break machines and that technicians do not always fix them, their differences of opinion about radio stations, and the fact that the technicians are there often enough that the receptionist knows all their first names, idiosyncratic spellings and all. However, under other circumstances, the technicians say these same issues can produce hostile confrontations instead of jokes.

For all users, in the interests of being thoroughly professional, the technicians will explain the circumstances of a machine that must be left unfixed, even for a lunch break. This is also a matter of respecting the user's responsibilities: dedicated operators and key operators are expected by their colleagues to know the status of their machines. These more involved users are also to be trusted with the technicians' tools and documentation; the technicians will leave them with the machine while going away for lunch or to get parts. This is also a guarantee for the proclaimed intent to return.

It is this combination of trust and mistrust, the sense of a common concern with different perspectives, that characterizes the relationship of customers and technicians. I have suggested that the technicians and the customers have a social contract, in which the technicians commit to fixing the customer's machine problems, while in return the customers let the technicians disrupt their place of business. It is not an entirely comfortable relationship for either party and exists as a complement to the formal, legal contract between the customers' corporation and the technicians'. It is, however, essential to the technicians' approach to the problem of defining what is going on with the machine.

6

Talking about Machines, and Bits Thereof . . .

The machines are the third party to the triangular relationship of service. In a sense, the machines create and partially determine the world of service, in that the relations between customers and technicians that have been described in the preceding chapters occur because of the presence and behavior of machines. The machines are the technicians' *raison d'être* and preoccupation, as well as occupation and sometimes passion, and so the technicians talk about them, continually, in a surprising variety of ways. In fact, this chapter is primarily about the ways in which technicians talk about machines and only somewhat about the machines themselves.

Just as sheep in a flock are individuals to their shepherd, so do specific machines appear in technicians' discourse as individuals with histories and known propensities for perverse or benign behavior. Machines are also depicted as the subject of procedures as the technicians do various things to them, trying procedures or changing procedures, all to fix the machines. It is in the context of diagnosis that details of the machines will appear in the technicians' conversation, since the details determine aspects of both problem and solution. Machines have parts, which have lives of their own until installed, at which point they may disappear and be subsumed into the whole, assuming they work. The machines are a presence, are subjects, are motivation for much of the action; if it seems that the whole machine never quite appears in the talk, perhaps that is because a whole machine would be complete and running and consequently not so interesting for the technicians.[1]

Technicians talk about machines in general, they talk about specific machines, and they talk about specific subsystems. Perhaps the most general talk about machines occurs just in the noticing of them. There is something inherently noticeable about machines for these technicians, a matter of being attuned to machines. They can tell you not only about their own machines at different customers' locations, but also about other machines from their company and from competing companies. If you show signs of understanding machines, as I did or tried to do, the technicians will make sure that you notice the machines too, as significant features in their world. The noticing includes an implicit or explicit comparison both of the machine's characteristics and of its reception by the customers.

Some of the general talk which appears to be about machines qua machines turns out to be about specific problems but divorced from context. For example, a technician flatly asserted to me that a given class of problem has only two causes, apparently a statement about these machines in general. However, it turned out that he had just been assigned a call on that sort of problem and knew that solutions based on those two causes had been tried without success, but those are the only two known causes. In a sense, this bald generic assertion is an invocation of the common wisdom as a spell in this strange, untracked but quite specific problem space. Similarly, the observation that a given component used to cause major problems and perhaps no longer does actually occurred in the context of having just tested it and having been spared again. Another apparently general comment, wondering if certain parts could fail when they never have, may have been less an effort to expand the technician's understanding of the machine than a hope for a new cause to explain present, immediate problems.

Talk that appears to be about generic machines, then, is often informed by one or more specific real-world examples. There is either a specific problem in mind, a specific customer situation, or

[1] Perhaps such a machine would appear in a customer's story.

a comparison to one or more specific behaviors that the technician knows personally. This same immediacy of reference can be seen when they discuss rumors of forthcoming machines. In one discussion, the technicians were quite explicit about the types of improvements that would address their problems with the extant machine. Other aspects of the new machine were either unknown or not of interest at the time.

The technicians have more to say when talking about specific machines or specific subsystems because of the increasing definition of the context. This specificity also seems to increase the enthusiasm of their discussion. When technicians talk about specific machines in their territories, it is clear that these machines are individuals. Their different histories, different patterns of use, and different social environments have given them each a distinct character for those who know. Given this individuality, the machines may be discussed with as much ellipsis as any mutual acquaintance. Generally, there is an identification of the machine with the customer, so a machine is commonly referred to by the name of the corporation that owns or leases it. For some corporations with multiple machines, any single machine may be known by the name of the person responsible for it, its street location, or some combination of these references. The model number will be combined with one of these for any machine other than the most common. In all cases, the machines are known by their social situations, and the lack of discussion following some mentions may be presumed to indicate that the name tells those listening everything they need or want to know about the machine and its situation at the moment.[2]

The specific local reference seems to characterize most talk about machines except when technicians differentiate the machines by class, speaking of the various models. Persons other than technicians often assume that all machines of a given model are essentially the same. It seems probable that technicians start with that assumption and watch it fragment as they learn more

[2] It is also true that there is no other convenient and memorable way to distinguish one machine from another. The corporation prefers serial numbers in official interactions, but technicians do not seem to find that a useful way to remember machines and their social contexts.

about individual machines. What is left, as a class identity for models the technicians work with, is primarily an approximation covering the disparate individual truths, although it would also include any ways in which the machines do, in fact, appear to be all the same to the technicians, the people who know them best. Class identity is clearest when technicians are discussing classes of machines fixed by other groups of technicians, where there is no local information to mar the unity of the class, but they may also compare their own flock as a class to other classes.

In such comparisons, technicians understand the capabilities and limitations of their own machines as types quite well. In particular, those with older machines are quite aware of their weaknesses and the greater difficulty of keeping them running well in comparison with newer machines. For the most part, they believe that the customers keep the older machines because they are cheaper; if this were not so, surely the customers would trade up to a newer, better, easier machine, would they not? Consequently, the discovery that a new machine with approximately the same capabilities costs about the same as an old, crotchety, legendarily difficult machine leaves the technicians baffled. It suggests that their work is being ignored or discounted, in that they must work much harder for the customer to get as much done with the old machine as they could with a new. It also suggests that someone in sales is missing the opportunity to sell the customer a newer and presumably more profitable machine to replace the old one.

It is true that the particular newer machine referred to above has had some problems, but these were dismissed with the wonderful distinction that they are "strictly parts failures, bad parts." There is an implicit claim here that there are no significant errors in design in this machine, just localized errors of execution. Thus, for this argument, the machine is seen as an assemblage, some few components of which have quality control problems. When these are cured, the machine will be whole again and fine. This ability to decompose the machine into its constituent parts, or compose the whole machine from them, is characteristic of the way technicians talk and maybe think about their machines; in this instance, it allows them to dismiss problematic behavior that is surely just as troubling to the customer as if it were admitted by the technicians to be a problem with the machine itself.

This belief in the fundamental integrity of the newer machine also permits the technicians to promote it in good faith as a replacement for the older one. Not only is the older machine difficult to repair, but there is no credit to be had for doing so, either from the company or from one's peers. There is a certain historic credit attached to having worked on it, in that this machine has the reputation of having created good technicians, but that is worth as much in the past tense as in the present. It is not, then, surprising that the technicians responsible for the old machines want them to go away.

Technicians are proud of their machines and the work the machines do for their customers. Those whom I studied took some pride in the fact that customers often preferred to use their machines rather than nominally much more powerful ones. They even took a perverse pride in the endurance of a machine run at volumes far greater than those it was designed to handle. However, while they are pleased when the virtues of their machines are recognized, they also recognize the costs that use, particularly excessive use, entails for the machine in terms of premature wear and for themselves in terms of higher parts budgets, increased number of calls, and diminished reputation.

In fact, the use of the machines causes some ambivalence for the technicians. On the one hand, the only reason for having the machines is to use them, and the only reason for having technicians is that when machines are used they break. The technicians get no joy from a machine that is underutilized. Such a situation has its own peculiar technical problems, and there is a sense of waste that the machine is not doing what it is good at doing. The technicians admire machines that are well designed, well constructed, and easy to use, and they like them to be used appropriately. On the other hand, while technicians have a certain amount of their identity invested in the machines, the customers do not, yet the customers have the machines. Figuratively speaking, the technicians' machines are in the hands of heathen. The technicians find it peculiar that the customers do not even know the proper language to use in describing the machines. They find it both peculiar and infuriating that the customers will not be bothered to learn how to use the machine properly but will persist in improvising their own methods, many of which do not work or even cause problems for the machine. The technicians do not mind the inevitable failures that result from use, but they object to those caused by ignorance and misuse, particularly what they see as wilful ignorance and misuse. It is not clear to the technicians that their social contract with the customers covers such abuse.

Using a machine will eventually cause it to break, and this both creates the technicians' job and detracts from their image of being in control. Using it beyond its designed performance will cause it to break far more often, and if this continues, the increased number of calls will make it difficult for the technicians to claim that they are controlling the situation rather than patching up the collapsing machine. Eventually technicians give up on the image of control; they continue to fix the machines that are being beaten to death but they no longer enjoy the work or take pride in it. It is not a role they want to play or a situation in which they wish to participate. In fact, the technicians seem to redefine the situation for themselves as not part of the normal world and not subject to their social contract. Therefore, their sense of professionalism does not require that they manage the situation but permits them merely to cope, much as they do when in someone else's territory. In such circumstances, they do believe that the machine should be replaced by one sufficiently more powerful that a normal relationship of customer, machine, and technician can be restored.

The technicians talk of such overuse as "machine stress," but it should be noted that the concept has several different meanings for the technicians, distinguished by different time parameters. There is the stress of continual use at a high level for a prolonged period of time or the stress of intense continuous use for a shorter period of time. Strategies for the care of the machine vary according to anticipated use and involve another sort of stress, the stress test. In stress testing, the technicians try to provoke weak components to fail or otherwise show themselves, and they change the stress tests according to the stresses anticipated in the machine's ordinary use. Ultimately, daily use is the stress standard, and a machine that cannot be provoked into displaying its problem is given back to the customer with instructions to run it until it breaks.

Much of technicians' talk about machines really involves keeping track of each other's movements and collecting the latest news about what is happening to their flock, and as such it is necessary business. This is not, however, the most interesting part of talking about machines for the technicians. What really holds their interest is a situation they do not understand. One such situation involved a failure of the system by which the corporation organizes the service world, dividing up the technicians and machines and assigning the latter to the former. A machine had been repeatedly reassigned until a time came when it was not assigned to any technician. This stray sheep had recently been rounded up and returned to someone's flock, thin, limping, fleece ratty and full of burrs. The absence of specific assignment usually results in the deterioration of a machine, since there is no individual technician responsible for maintenance and updating, and the technician newly in charge of the stray was spending long hours rebuilding its health. The work required by the machine was no surprise to the technicians; the curiosity was that the machine had been lost from the system on which they depend to help them keep the world under control. There is also an implicit assumption in technicians' talk about machines that machines are normally located in society in two ways, in the customers' sites and in the technicians' charge; slipping from either location makes a machine both an anomaly and a threat to the social order of the machine world.

Acquaintance with particular machines can guide repair strategies in more successful paths as well. A technician frustrated by a recently inherited machine proposed changing every instance of a component in that machine because one was causing problems. A more experienced technician discovered which machine was being discussed and revealed that this had already been done, suggesting instead that only the offending component be replaced. Conversation among several technicians about this revealed their belief that such mass replacements are rarely justified; extensive disassembly, reassembly, and readjustment are apt to create as many problems as they solve. There is a sense of existential peril revealed here in the recognition that they will make mistakes and that they should therefore limit their interventions with the machine. Making a mistake while fixing an actual problem is more readily excused than making one during a global replacement of parts not yet gone bad, work whose value is not likely to be recognized by the customer anyway and whose failure has the real possibility of turning an ordinary service call into an extended nightmare. Such drastic measures are seen by the technicians as occasionally necessary, but undertaking them without very careful consideration is wanton flirtation with disaster, with a high probability of losing control of the situation, of the state of the machine, of one's image, and of one's status with the customer. In short, of losing everything.

The technicians do not enjoy work with risks on that scale; they prefer working farther from the edge of disaster. Work that permits them to show off their skill with little concomitant risk arouses considerable enthusiasm. One group at lunch got particularly involved in the discussion of a repair that had been developed in the field to cope with mechanical wear until improved components could be designed and distributed.[3] The repair is nearly permanent and creates some difficulty when it is necessary to disassemble the repaired mechanism in order to service other components. The technicians like this fix for several reasons. The problem is most often found in the presence of the complaining customer, producing a very showy diagnosis which underlines the skill and knowledge of the technician. The fix is aesthetically pleasing in that it is both very effective and very economical; it requires minimal disassembly, reassembly, and adjustment, uses scrap material for a shim, repairs the mechanism to closer tolerances than those of manufacture, and lasts. The closer tolerances do make its eventual disassembly harder but not impossible and so enable a subsequent display of skill. Reassembling and disassembling the shimmed mechanism requires a modest amount of brute force but with little risk for other components; the principal problem with the fix was that some technicians were overly enthusiastic and shimmed the mechanism so tightly as to make it very

---

[3] This is the repair that Frank did in the fifth vignette of Chapter 2.

difficult to disassemble. The fix is also theirs: it was developed by technicians in the field to solve what can be seen as an engineering failure. It embodies their skill and its value for all to see and so constitutes evidence to support their claim to skilled status.

Another way in which the machines capture the technicians' interest is through behavior which, while not compromising performance, does not fit with the technicians' understanding. A machine performing an automatic self-adjustment procedure was not displaying the information expected in that procedure. The technician working on the machine interrupted the process several times to try to find out what the machine was doing. Eventually it was perceived that the machine was performing the procedure correctly but not giving the correct display; this was immediately attributed to a recent exchange of electronic subsystems. Subsequently the team's technical specialist told us that this behavior indicates the presence of a subsystem elsewhere in the machine used by engineering for development work; these were not supposed to appear in the field, but some did anyway.

The behavior continued to bother the technician responsible for the machine, since the subsystem indicated by the specialist had actually been in the machine for over a year without ever displaying this behavior, while the subsystem that had recently been exchanged is the one that drives the display. The technician responsible made a point of telling other members of his subteam about this peculiar behavior so that they would be prepared if they should take a call on the machine.[4] Later, he told this tale to another highly regarded technician as a riddle, presenting the odd behavior without any explanation, but the other technician immediately came up with the same explanation as the technical specialist. This still failed to satisfy, since it did not explain the previous good behavior or the sudden appearance of the anomaly after changing a nominally unrelated subsystem.

The technician's discontent is profoundly qualified, however, by the fact that the machine is running properly for the customers. This ultimately satisfactory behavior means that remaining anomalies cannot be terribly important. They do still irritate, in the sense that a technician's ability to cope with any possible machine

---

[4] This appears in the breakfast conversation in the first vignette in Chapter 2.

behavior depends on understanding how any such behavior is produced. Presently benign behavior which is not understood may appear later in less benign manifestations. Accordingly, the concern lingers at a low level, the sense that something is not well understood balanced against the currently acceptable behavior of the machine.

The technicians' relationship with the machines is partially kinesthetic, knowing how the machine should feel. As described in the fifth vignette, the diagnosis leading to the shimming mentioned above involves wiggling a shaft; the presence of excessive play indicates the need for repair, but knowing how much play is too much requires a sensitivity to the feel of the mechanism. Other adjustments are made on the basis of feel, or the need for adjustment may be judged by feeling the mechanism work.

Related to this point is the technicians' use of the sounds of the machine. In several diagnoses that I observed, the noises produced at different stages of the process proved to be an invaluable guide to what is happening or not happening. One set of sounds indicates where the problem occurs, another indicates a particular sort of problem, and yet another indicates that the controlling logic has just crashed. In older machines, the succession of noises narrates to the experienced ear the progress of the operation, and should it fail, the last noises suggest where to look for the problem. Perhaps more obvious are the sounds of mechanical distress, as mechanisms bind, bearings go bad and squeal, or pins slip out to stop the rotation of a shaft completely while an overzealous drive belt thumps away, skipping one tooth at a time. One of the few complaints the technicians ever expressed about a customer site was that one was so noisy that they could not hear the sounds made by their machine. In fact, the noise level was such as to constitute physical abuse, but what concerned the technicians was that they could not hear the mechanism operating.

Finally, the machines are both perverse and fascinating. Earlier models featured both fires and explosions, and the technicians speak with a fond pride of the labor involved in recovering from such disasters. Catastrophes resulting from oversight are described with the same pride as part of the process of becoming a "real technician." The machines can be merely difficult, but the technicians show no resentment as they describe the hours of troubleshooting necessary to make the elusive connection between the inconclusive behavior of the machine and the crucial failure. Indeed, how could they resent the machines, for such a machine is a worthy opponent, partner, other. It is in terms of these machines that the technicians construct their own selves, both hero and fool, both with pride in coping with the machines' perversity and keeping control of the situation and with humility in acknowledging their lesser failures, while praying never to have a situation that simply cannot be handled. The nature of the machines also governs the social relationships of technicians and customers and technicians and managers, and the technicians value their autonomy and the use of their skills. If the machines behaved, the technicians could not be technicians.

Spare parts are the constituent components of the machine, and yet they are also a concern in their own right. The technicians alternately talk and presumably think of machines as whole and as composed of parts; conversely, they also talk and think of parts by themselves and as belonging to machines. The persona of the machine, however, is more than the sum of its parts, and the machine retains its identity despite extensive transplants. It is unclear when the parts actually become part of the machine, possibly when the machine runs again, free of the trouble that provoked the transplant. Parts are problematic in their own right, however, and the principal issues are availability and reliability.

Technicians carry a supply of parts in their car, which they replenish weekly. This trunk inventory, as it is called, is intended to be adequate for all routine service and repairs. Most heavily represented are those parts replaced routinely or which fail most often; parts that fail infrequently or are very expensive are not often included. The inventory is constrained by a parts budget, which is one of the criteria by which the technicians' performance is judged, so they cannot simply stock everything. The technicians get their parts from a district warehouse, which works under roughly the same constraints as the technicians and so does not stock large numbers of the more expensive or rarely used parts. This could mean that a part is simply not available when needed, which would necessitate an Emergency Order, which may in turn be frustrated because the Regional Distribution Center, also subject to the same constraints, may be out of its limited stock as well. This limited availability and the use of parts budgets to judge performance are the only evidence of a presumably comprehensive corporate spares strategy to be seen from the field.

Given this potential frustration, the technicians have evolved a variety of strategies for keeping the necessary parts available. As we have seen, one strategy is to share inventories, setting up a scattered distribution of the more expensive parts, and the groups involved seem to be roughly equivalent to those exchanging information. A technician in need asks around, often at a common meeting point like lunch or the Parts Drop, to see who has the necessary part. A technician appearing at lunch with an odd problem will be offered those parts that seem appropriate to the problem. Another resource is the team's technical specialist, who maintains an inventory of unusual or expensive parts that individual technicians are not expected to stock. Like any of the other resources, this inventory is limited and may be out of the desired item.

The final recourse is often a machine at the branch office, which may be robbed to keep a customer in business. Often there is a specific machine, known in this district as the Hangar Queen or Queenie, which is rarely or never complete, as one piece or another is always being borrowed for a desperate customer. This approach was said to be understood by the corporation but not officially condoned. There is some doubt about the reliability of the parts remaining on Queenie. Often the parts have been removed from a machine in trouble, but the replacement parts from Queenie did not fix the problem. This means that the parts removed were not the problem, but they were part of a troubled system, whose difficulties may have stemmed from the interaction of various subsystems rather than from a single source. It cannot be said definitely that the parts in Queenie are working perfectly. Nevertheless, when technicians get sufficiently desperate, they will still take parts from Queenie. In such instances, the parts do not merge with the repaired machine, at least not immediately, but remain identified as Queenie's parts and so are considered suspect, on probation, and perhaps only provisionally there until a new component can be obtained.

Technicians' concern with the problematic nature of the supply of parts has to do with both the reality and the appearance of control over their situation. The machine cannot be fixed without parts, and the customer will notice if the machine is not fixed, resulting in damage to the technician's reputation as well as the machine's. Furthermore, if a technician has to leave in the midst of a repair to get a part, the disruption of a broken call is more noticeable than additional time spent working on the machine. This may be motivation to try to combine trips for parts with trips for lunch, to reduce the commotion of departure and return and to hide the break under a more innocent rubric. The longer the call, the greater the embarrassment; breaking an already long call rubs salt in the wounds. A particularly heartbreaking example occurred during an all-day overhaul of a long-neglected machine. Well after normal business hours, the two technicians working on the machine discovered that it needed an ordinary part that they normally stock, but neither technician had one. At that hour, there was no hope of getting the part from someone else, so one of the technicians would have to return in the morning, nullifying much of the effect of their after-hours heroism by disrupting the customer's routine once more.

The other major issue with parts is reliability: both their longevity in use and the percentage that are bad when received. The nature of the machine dictates that certain components will not last very long, at least as currently manufactured. The response by the technicians to the failure of such a part varies. One could decide that the first part to fail indicates that all such parts should be replaced. As indicated in the discussion of the machine-as-a-whole, this procedure risks creating more problems than it fixes, particularly if changing the part involves extensive disassembly, reassembly, and adjustment. On the other hand, one technical specialist commented that "a switch is a nit" and indicated that he would cheerfully change all the switches in the machine rather than worry about whether any one of them is going bad, but then this is an exceedingly straightforward task with little danger of complicating matters. The technicians have also developed some modifications to improve the performance of known problem parts; this knowledge is shared through the community in the same way as new diagnostic knowledge, through war stories.

The other reliability issue is whether or not new parts are functional. Some of them are bad to begin with, and it is a quick trip from the box to being tested to the trash can (or, for expensive pieces, back to the Parts Drop to be sent off to be fixed again). Others will work for a short period of time and then fail well before they should. Once this happens, the world becomes much more problematic: one can no longer assume that a part of the machine will work simply because the part is relatively new. This effect is increased for problems where diagnosis is not certain because there are no reliable tests to isolate the cause of the problem, merely a number of known possible causes for that problem. One can no longer reduce that number because a given part has just been replaced.

Parts are not always scarce, suspect, or otherwise worrisome; there are more positive aspects to the parts situation as well. New parts are always greeted with enthusiasm; the promise that they will solve the problems they are designed to address is initially accepted at face value. A new part designed to replace the one the technicians had had to shim in the field was thought to be a great improvement. The fact that some previous new parts have not been perfect solutions does not seem to diminish the enthusiasm for still newer ones. Each new part might improve the situation; if not, the technicians will cope, exactly as they have been doing.

As we have seen, parts modified by engineering sometimes turn up in the field; the technicians are wary of machines with such components. Such parts are intended to perform normally most of the time. However, they may omit some normal functions in order to accommodate special functions for the engineers' purposes, and therein lies the rub. The omitted functions may seem like a problem to the technicians encountering them in the field, until everyone knows that a certain machine has some engineering parts and will not behave as one would expect. What is worse is that the engineering functions are completely unknown, and the presence of an unknown element in the machine complicates diagnosis. A technician could never know whether problematic behavior is caused by the unknown portions of the machine or not; the only recourse would be to change those parts first to see if the problem disappears.

Swapping parts is, in fact, a diagnostic strategy of fairly late resort. If one cannot reach an understanding of a problem, one can trade parts until the problem goes away; however, one does not then know exactly what was wrong, and so is unprepared for another instance of the problem. Changing the parts changes the conditions that one is testing, so the disappearance of the problem is not necessarily attributable to the new parts. Changing large numbers of parts is even less certain, in that many of the most difficult problems seem to be caused by the interactions of parts not in themselves bad; a new cast of players changes the interactions but does not reveal the source of the problem. Thus, swapping is not a desirable strategy but is certainly preferred to not fixing the machine, which is unacceptable. Parts swapping is even prescribed by the documentation for fixing some problems where the corporation did not or could not give the technicians the information or equipment necessary for diagnosis. It is effective, it is sometimes more efficient than diagnosis, but it leaves some discomfort in that one never quite knows what the problem was, only that it is gone for the moment.

7

The Work of Service

If the work of service, the job of technicians, is the maintenance of a three-cornered relationship among the technicians, their customers, and their machines, the crucial issue is knowing what to do. A service call is occasioned by a problem in the relationship between the customer and the machine, and neither is articulate in the terms used by technicians to talk about their work. Customers talk about machines in their terms and those they have learned from technicians, while machines communicate through behavior, error codes, and machine logs; from all of these, the technicians must construct an understanding of the situation that permits them to resolve it.

The stereotypical view is that service is about fixing identical broken machines, and the technicians do indeed work on the machines. They must diagnose and repair the problems of the machines, as well as maintain and adjust them. In all of these activities, and perhaps most critically in diagnosis, the technicians must understand the machines. Understanding the problem determines what is to be done about it, but that understanding is created from an assortment of information that does not necessarily point to a single diagnosis. The practice of diagnosis is done through narrative, and both diagnosis and process are preserved and circulated among the technicians through war stories, anecdotes of their experiences.

This, however, is the view of an ethnographer observing diagnosis as it is done. The corporation has a different view of the work, including diagnosis, the gist of which is that the technician needs to understand little more than how to follow the directive documentation furnished by the corporation. This view, in turn, affects the information provided in the documentation to the technicians and so affects the doing of the work, although not always as intended. In order to understand how the technicians work with the machines, in particular diagnosing their ills, one must understand how the documentation is done and how it is intended to be used.

The Use of the Service Documentation

Latour (1986, 1988) tells us that machines prescribe human behavior, forcing us to do certain things to use the machine or other things to accomplish our ends without using the machine. This is part of his argument that machines participate in human society to such an extent that neither technology nor society can truly be considered apart from the other. The machines never participate with intent, however, and the humans at least some of the time do. Machines represent the intent of their designers, and so are an extension of human interaction with humans. Furthermore, intentions and their results must be considered in their social settings; the machines must be seen simultaneously as products of the social context of their design and production and as participants in the goals of the users. Madelaine Akrich (1992) maintains that the reality of the machine is not in the machine itself, in its designers' intentions, or in its users' intentions, but in all three at once, particularly as they intersect in the situation of use. One of the ways in which Latour says the machines prescribe human behavior is through owner's manuals. Consider, then, the issue of service documentation.

The usual conception of a service manual is that it contains a mixture of descriptive information about the machine that is the subject of the manual, such as mechanical drawings or electrical schematics, and some instructions regarding the maintenance and repair of the machine. The presumption is that users of the manual get information about the machine from the manual, which they use to think about any problematic behavior of the machine and to deduce the source of its problems. However, documentation is not just a representation of the machine and its prescriptions but must be regarded as a mechanism in and of itself. A service manual is a device which someone constructs to convey information to someone else, and choices of inclusion and exclusion significantly constrain what can be done with the manual.

The corporation chose to use directive documentation for the principal machine serviced by the technicians in this study; this is a style of service manual that purports to instruct the technician in every development of the service call. In directive documentation, the information is selected and arranged according to the documentation designers' projection of what will be necessary for the tasks that the technicians are intended to perform. The designers' choices are constrained in two ways: first, by their own source of information about the machine, which is the engineering group responsible for its design and production; and second, by the service organization's policies about how service is to be done. Thus, the design of this device for conveying information is done by one group with information from another group according to policies from yet another group, and the policy input has the effect of changing the service manual from an information device to one that also attempts to determine how the work will be done.

This directive documentation is designed not to provide information for thinking about the machine and its problems but to direct the technician to the solution through a minimal decision tree. The directions in this documentation are intended to prescribe the technician's behavior from arrival at the customer site until departure. The premise is that a careful following of the prescriptions from beginning to end will lead to the resolution of problems more quickly than could be accomplished by the technicians reasoning from their understanding of the machine. A necessary corollary to this premise is the belief that all significant problems can be anticipated and their solutions prescribed in the documentation. Such directive documentation may omit information that would contribute to understanding the problem and provide only the information believed necessary for following the instructions. The success of this scheme for providing service clearly depends on the success of the documentation designer in correctly anticipating and providing for the troubles that actually occur in the field. Success as directive documentation also depends on the users, the technicians, understanding how the documentation is intended to be used and making the appropriate use of it.

Directive documentation belongs to the scientific management tradition of attempting to rationalize the work process (Braverman 1974). The basic premise of scientific management is that one can reduce the best way to do a given job to a set of instructions and give those instructions to someone who does not know how to do it independently but who will then be able to do the job by following the instructions. In this way, management gets control over their employees, through control of the knowledge necessary to do the job, and can hire cheaper employees, since they do not need skilled labor. The whole enterprise rests on the ability to define the best way to do the job and then to provide adequate instructions. Harold Garfinkel (1967) and Suchman (1987) have convincingly demonstrated that self-contained instructions are not possible, and I contend that the knowledge relevant to the job of diagnosis cannot be precisely defined.

The directive documentation issued to the technicians in this study contains a set of prescriptions describing how to maintain the machine's health, cure its problems, and generally coerce it to function. The diagnostic procedures prescribe a series of tests, with each action defined in considerable detail, and each branching condition presented as a simple Yes/No choice. However, the actual question to which one is responding Yes or No may be extremely convoluted. The corporation requires that documentation be written in a form suitable for simple machine translation, severely limiting the permitted vocabulary. It also requires that documentation be produced using an automatic formatting system which limits the arrangement of branches of the decision tree by restricting the number of available tabs or indentations. These requirements of machine translation and of document formatting both produce questions to which the answer is not obviously Yes or No, even with the results of the test definitively in hand. Solutions to problems are presented as branches to similarly directive, if somewhat clearer, repair procedures. No rationale is offered; the explicit purpose of the tests and the interpretation of the results are both known only to the designers of the documentation.

At the time of my fieldwork, however, the manuals combined the directive procedures with simplified schematic diagrams, showing the interconnection of all systems, digital, other electrical, electromechanical, mechanical, and pneumatic. The training material included a book called the "Principles of Operation," describing how the various systems function in considerable detail. Such a book in combination with schematics and some directions for adjustment or replacement of parts constituted the documentation for earlier generations of these machines. In working on the earlier machines, the technicians used the information in the documentation and the information they gleaned from the problematic machine to diagnose its failures or maladjustments through their own reasoning. At the time of this study, the elements of the traditional documentation stood in an ill-defined coexistence with the directive prescriptions which were intended to dictate solutions to the service problems in the field.

The technicians, however, use the documents in pursuit of their own goals, and these are only somewhat the same as those of the designers of the documentation. A technician's primary goal is to keep the customer happy, and this includes but is not limited to fixing the machine as necessary. An important component of this goal is keeping the customer assured that the situation is under control, which requires being able to tell what the machine is doing and being able to say when it has been fixed and what has been fixed. The customer must know that the technician has repaired the machine in order to feel confident that the machine will be repaired in the future. Accordingly, a system that fixes the machine without either customer or technician knowing how or why is unlikely to be acceptable. Consequently, when the technicians use the directive documentation, they try to determine the purpose of the various tests, to understand what the documentation is testing, to know what they are doing.

This is not done just so they can reassure the customer; the technicians are also developing their understanding for future problems. The most common machine failures, which are the ones most likely to be correctly anticipated by engineers and documentation designers, quickly become routine for the experienced technician and no longer require documentation. The unusual, rare, exotic failure modes are much harder to anticipate and may not occur with sufficient frequency to justify more efforts to anticipate them. However, for the customer whose valued machine is not functioning, the rarity of the failure is no consolation. The technician who is responsible for the machine must still fix it, both to preserve the social contract with the customer, who expects the technician to be able to fix the machine, and possibly to preserve the legal contract between the two corporations. Customers may cancel their leases if dissatisfied with the service support they receive. Accordingly, the technicians must prepare to solve new and unanticipated problems, which requires them to develop as comprehensive an understanding of the machine as possible. When technicians use the documentation, they contrast their analysis of what the documentation is trying to do with their own analysis of what might be wrong with the machine. They pursue those paths in the documentation which seem consonant with their hypotheses. It is almost certainly true that this is less efficient than simply following the documentation for those problems which the documentation solves, but it may well help to develop the skills necessary to solve problems not anticipated by the documentation.

Thus, there are social and technical reasons why the technicians should have a good understanding of the machines and the ability to verbalize it. There are also both technical and policy reasons why the documentation may not provide all the information they need. Before we turn to how diagnosis is actually done, we will consider the various ways the technicians use the service documentation within these constraints, as well as the ways they talk about using the documentation.

I suggested at the beginning of this section that the service documentation should be viewed as a mechanism designed both to convey information about the machine which it represents and to shape the way service work is done to that machine. However, where other machines can prescribe human behavior in their use, although not necessarily as intended by their designers, the mechanism of documentation is severely limited in its prescriptive ability. It is composed of representations, which inherently afford multiple interpretations and uses, and instructions, which require interpretation by their users in the context of their application. As Suchman writes, following Garfinkel, "Indexicality of instructions means that an instruction's significance with respect to action does not inhere in the instruction, but must be found by the instruction follower with reference to the situation of its use" (Suchman 1987, p. 61). The documentation cannot constrain its users except through omission of information, and questions of what to include and what to omit are, in fact, part of its design as a conveyor of information. Its intent to shape the performance of service remains a desire and never achieves the force of Latour's concept of prescription (1988), which would enforce the desired behavior.

This is perhaps fortunate, since many of the scripts created for the desired behavior seem to be flawed. The technicians' talk about using the service documentation is full of cautions about the perils of following the diagnostic procedures. Some of them are said to be circular; some are easily misread. Even those technicians who profess the strongest attachment to the diagnostic procedures warn that, without understanding the intent of each procedure, one can easily make the wrong choice and get hopelessly lost. The possibility of making wrong choices was supposed to be precluded by the design of the procedures; the technicians' experience indicates that this failed, and the failure has made them wary of the documentation. Even the most skeptical technicians concede that the documentation does solve certain problems quite well; they insist that if the documentation offers a solution to a given set of symptoms, it cannot be discounted without trial. One suspects, however, that this is not the level of credibility hoped for by the designers. There is also a certain amount of resentment of the diagnostic procedures; as one technician told me, technicians like to think they have more on the ball than just following directions.

At the same time, the procedures are viewed as providing a

systematic approach to the diagnosis and repair of machine problems, and being systematic is highly valued among the technicians. Their greatest scorn is reserved for technicians who somehow never get to the problem that triggered the service call. The general opinion seems to be that if a technician does not know what to do, they should follow the procedures. It is conceded that this will require more work; in the interest of thoroughness, the procedures are thought to require more than may be strictly necessary. The technicians feel, however, that shortcuts are only warranted if the technician can be sure that the omitted procedures are irrelevant to the problem at hand. The team's technical specialist had prepared a guide to the shortcut process, listing certain common procedures whose relevance could easily be seen without performing all of the official preliminary procedures.

This slightly subversive act probably typifies the expressed attitudes of the technicians toward the documentation; they grant the documentation some utility while denying it complete credibility. These attitudes are further complicated by the technicians' perception that they must project an image of competent practice and the fact that the corporation requires use of the documentation. The former dictates that they systematically try all possible approaches to a recalcitrant problem, and the latter grants a form of immunity to blame should the problem prove intractable. That is, in providing directive documentation, the corporation is assuming responsibility for solving the machine's problems, and in the eyes of the corporation, technicians are only responsible for failure to fix a machine if they have not used the documentation. However, while the technicians are quite willing to let the corporation assume any blame, their own image of themselves requires that they solve the problems if at all possible. This means pursuing all paths including the documentation, which does, after all, fix some problems; any new problem might actually be one of them. The technicians are quite philosophical about the shortcomings of the documentation, saying that the machine is far too complex to anticipate correctly all of its possible failures. They view the documentation as a useful resource to consult when their own expertise cannot solve the machine's problem, and they do so.

It is important in this discussion to remember that the documentation actually includes different components in different styles, combining directive diagnostic and repair procedures with schematics and a functional description of the machine. The technicians choose from this array of documentation; I never saw some books, such as "The Principles of Operation," in the field at all. The technicians primarily use two resources: a book that combines the directive diagnostic procedures and the schematics, and a book or set of microfiche cards that describes the repair procedures and lists part numbers.

The consistent theme to the use of documentation is that the technicians always turn to it when they do not understand the state of the machine and ignore it when they do. It seems clear, however, from the ways in which the technicians use the documentation, that they are as aware of its shortcomings in practice as they are in conversation. The majority of diagnostic procedures proved in the cases I observed to be an unreliable crutch, since the procedures only work when there is a symptom linked to a procedure which solves the problem producing the symptom, and this is not always the case. Several of the calls in which the use of the diagnostic procedures was observed were return calls to machines whose problems had persisted after earlier efforts to fix them; this persistence clearly indicated to the technicians that their symptoms were produced by causes other than those known to the documentation. Under these circumstances, when the technicians followed the procedures, they interpreted them, reading all the branches and all the proposed solutions to try to understand what the procedure was intended to do. The procedures were followed until they indicated the replacement of parts already replaced. This conclusion was generally rejected, although the known fallibility of replacement parts required the technicians at least to consider the possibility that the procedure was correct. There is also a procedure for those problems known to produce symptoms normally associated with other problems; this is the procedure to which most technicians turn after concluding that the answer provided by the initial procedure is wrong. If this provides no help, the technicians then turn to the schematics of the suspect area of the machine, using that information together with their interpretation of the intent of the diagnostic procedures and all other information about the state of the machine to do their own diagnosis. Perhaps half of the documentation use that I observed was use of the schematics, and most of this occurred after an attempt to solve the problem with the diagnostic procedures.

There are variations in the use of the diagnostic procedures. Some technicians follow them to structure their work, to ensure that systematic quality which technicians believe is vital to competent practice. This method was generally successful and provided a good illustration of the interpretive and technical skills required by the procedures. Other technicians browse. This occurred most often when the problematic machine was either giving no information or no valid information. In these situations, technicians were observed to browse through large numbers of procedures related in any way to what was known of the problem, looking for something that seemed to address the situation, either as solution or as inspiration to integrate the fragments of information about the machine into a coherent representation of its troubled state, which could then be solved. Finally, the most successful use of the procedures was for problems anticipated by the documentation which had not often been encountered by the technicians; they worked perfectly.

In summary, then, the technicians use the documentation routinely, but rarely with the blind faith intended by its designers. The technicians rely most on the schematics, believing their portrayal of the interconnections of the machine to be generally accurate. They approach the diagnostic procedures warily, knowing well that the full spectrum of machine misbehavior has not been and probably cannot be anticipated, but believing that the testing procedures in the diagnostics can be useful if the technicians can interpret both questions and results correctly. The repair procedures are seen as suggesting a sequence of operations, if the technicians have not already developed an easier way to do the same thing. All the documentation is seen as fallible; the task of maintaining perfectly accurate documentation for a complex machine through multiple revision levels of both machine and documentation is seen as improbable at best.

DIAGNOSIS

It is clear that directive documentation has not replaced the technicians' own diagnoses. Diagnosis, repair, and maintenance are the defining tasks of the technicians' job. Repair and maintenance are not in any sense unskilled work, and the technicians eagerly discuss refinements of their practice. However, the most serious technical challenge is to be able to learn what is wrong with a broken machine and thus how to repair it. It is also the greatest social challenge, since the very existence of a service call is based on a customer's perception that something is wrong with a machine, and the customers often control the critical facts enabling diagnosis.

One should think of diagnosis in terms of Suchman's proposition that "in the course of situated action, representation occurs when otherwise transparent activity becomes in some way problematic" (Suchman 1987, p. 50). The argument is that when an activity is proceeding smoothly, the equipment and skills used to perform the activity and one's knowledge of equipment, skill, and the activity itself are all invisible and unthought-of. One constructs a formulation of one's understanding of the situation when one needs to think about the activity, the equipment, the skill, or one's knowledge about them, because they have become in some way problematic. "Becoming problematic" may mean that the activity has been disrupted by failure, that one is perplexed about how to proceed, or merely that someone else has inquired about the activity, requiring an explanation.

The subject of diagnosis is usually a situation in which the customer has concluded that the machine has failed. The customer then must represent this situation in requesting a service call, and these formulations vary according to the customer's understanding. The most basic is the statement, "It's broken"; well-trained customers learn to provide more detailed representations of the problem, and the operators who receive calls for service are experienced in eliciting necessary detail. The technician assigned the call receives a representation of the problem which is the joint product of customer and operator.

The technician's task in diagnosis is to create a representation of the problematic situation that is sufficiently complete to indicate a course of repair. If the problem is known and recognized, the recognition may constitute sufficient understanding of the situation for the technician to resolve it. In this situation, no representation will be formulated unless someone else asks about the problem. If the problem is not recognized, however, an analysis must be done using information from a variety of sources, and the most difficult diagnoses are those for which none of the information sources provides a clear answer.

There is a class of problems for which the machine provides no direct diagnostic information. In such cases, diagnosis is accomplished by piecing together clues gleaned from the machine and the customers. These clues do not clearly indicate a specific cause for the problem. Their significance is in what they show about patterns of machine behavior; if interpreted correctly, and various interpretations are possible, such clues may suggest further areas to investigate which may produce a definite cause. Some of these diagnoses actually fall in the "known-and-recognized" class, in which the connection between clue and problematic behavior was established some time earlier and is now well known throughout the community of technicians.

For an example of a recognized problem, consider the transcript of the diagnosis discussed in the fifth vignette. On this particular service call, the customer complaint was that there were jams in a particular area of the machine. This was a relatively credible report, since the machine had not been serviced in a month and a half, so we went off to the customer site assuming that the machine was merely dirty. On our arrival, the technician asked to see the person who had been having trouble and the specific job that was causing problems, if possible. Two users who had experienced problems with the machine appeared, and this conversation occurred:

First User: I was having the problem with the feeder. Uh, I didn't bring my originals with me, but I was telling Richie [*the manager*], that they were flat, new originals, never had staples in 'em or anything. 't would feed 'em through [*opens machine cover*] and one would get caught right in here. . . .

Second User: That's where mine get caught.

First User: [*Closes machine*] and then I would have two layin' on the glass.

Technician: Two on the glass? OK.

First User: Two on the glass.

Technician: Thanks [*sounding very settled, decided, knows what's wrong*]. That's big input.

Second User: She was having problems with, uh, double-sided the other day.

Technician: Two-sided original?

First User: Yeah.

Second User: Two-sided.

First User: Yeah, it would make it through, it would go through on the first side, but then on the second time it would catch right in here. On top. . . . [*unintelligible*]

Technician: OK. As soon as you hear that extra noise, where it's clunkety-clunk, clunkety-clunk, as it's turning over. . . .

Users: Uh-huh, yeah.

Technician: OK. Thanks.

First User: OK?

Technician: Yeah. Thanks. You've told us a lot.

First User: OK.

Technician: Narrowed her down in a hurry [*door closes behind users*].

Ethnographer: Ah. What was she pointing to, the reversing roll in there?

Technician: Mm-hmm [*sound of machine covers*]. This is, well, this actually drives the roll.

Ethnographer: Ah.

Technician: [*Sounds of machine parts being wiggled.*] OK, I know what's wrong. That play is not supposed to be there [*sounds of machine parts being wiggled, loudly, demonstratively, for even an ethnographer to appreciate*], it's a common, it's one of the first things we check for.

Ethnographer: And where is it misadjusted? Where is the adjustment?

Technician: It's. . . . What it is, it's a D-shaft in back, and what it does, it's got a plastic bearing that drives it.

Ethnographer: Uh-huh.
Technician: OK, and the flat on the D-shaft, OK, wears out that flat on the bear—, on the pulley, the gear, and gradually enlarges it.

One of the striking points about this situation is the richness and variety of information sources available for diagnosis; these sources include the customers and the physical state of the machine. Also available, but not used here, are the error logs which the machine creates of its own events, the history of the machine written in the logbook by the various technicians who have worked on it, the service documentation for the machine, and the community memory of the technicians, in which they preserve and circulate their hard-won knowledge of machine arcana, usually in the form of war stories. Nor is the technician's own memory of these stories all that is available; other technicians called for purposes of consultation will bring their own recollections to bear, and a good memory will make one a popular resource.

However, the customer who experienced the problem is the first choice for information. The complexity of the machine makes it entirely possible that the problem will not recur in testing. In the absence of someone who can explain what happened, one then finds the paradoxical situation of a technician attempting to "duplicate the problem," that is, to break an apparently functioning machine. That does not happen in this case. The information from the customers and a quick check of the machine complete the diagnosis. No representation of the problem occurs until the ethnographer wants an explanation, and the difficulty of producing one suggests the extent to which the information was embodied in the technician.

Before discussing that production, there are some features to emphasize about the diagnosis. The first point to notice is that these customers have been well socialized in the ways of the machine. This is done by the technicians, persuading the customers to notice significant events about the machine and to talk about them in appropriate language, that is, the technicians' language. These customers have noticed the critical details about the originals, where in the process trouble occurs, and where the machine leaves paper. They have learned to create lucid representations of the situation, useful to the technician in the task of creating an account of the situation so it can be fixed.

This socialization of the customers, where possible, is a large part of the social work of service, and this episode contains one of my favorite examples of how far it can go. Consider the moment when the customer says: "Yeah, it would make it through, it would go through on the first side, but then on the second time it would catch right in here. On top. . . ." The technician replies, "OK. As soon as you hear that extra noise, where it's clunkety-clunk, clunkety-clunk, as it's turning over. . . ." They recognize the noise. It is true that this particular noise is quite distinctive, it can be said to afford notice, and anyone who had used the machine extensively in this mode would be likely to recognize it. This seems to me an extreme form of socializing users to machines, in which the technician is imitating machine noises and the customers are contrasting his noises with their own experience of machine noises. It is also an interesting strategy to conjure up a specific machine event and represent it for consideration in the context of diagnosis.

With this critical bit of information, the technician has a tentative diagnosis which is confirmed by opening a cover and wiggling a shaft, feeling for excessive play. The feel of the shaft confirms the diagnosis. This kinesthetic information is immediately tellable, and the technician says, "That play is not supposed to be there." This is followed by a demonstrative wiggle of the shaft so that I can appreciate the free play whereof he speaks. Then the technician explains that this is a very common problem, which perhaps explains the reduction of this diagnosis to the kinesthetic awareness of the proper feel of a shaft when wiggled.

The next part of this exchange shows the difficulty of representing the knowledge implicit in such an automatic diagnosis:

Ethnographer: And where is it misadjusted? Where is the adjustment?
Technician: It's. . . . What it is, it's a D-shaft in back, and what it does, it's got a plastic bearing that drives it.
Ethnographer: Uh-huh.

Technician: OK, and the flat on the D-shaft, OK, wears out that flat on the bear—, on the pulley, the gear, and gradually enlarges it.

This information, being less important than the critical presence of free play, which has a well-known repair, is clearly not so available for representation, is not immediately tellable, but emerges in a pair of difficult utterances, with many repairs. This information emerges, however, in its apparent order of importance. The first point is that the shaft ends in a D-section, which is instrumental in the development of the free play. The second part of the utterance is that "it's got a plastic bearing that drives it." It is *not* a bearing, but this is much less important than the fact that it *is* plastic. The second utterance elaborates how the flat of the (steel) D-shaft wears out the (plastic) flat in the matching item, which is still relatively unimportant but is recognized as not a bearing. In three tries, the technician produces the name "gear" for this item, and the representation of the mechanism is complete. The technician then describes (on the tape and in the transcript, not here) how this worn mechanism, which still actually performs its function in some sense, nevertheless shuts down the machine by failing to meet its very strict timing requirements.

But this was an easy diagnosis. For difficult diagnoses, the representation of the problem takes the form of a narration of the diagnostic process, a verbal consideration of what is known or thought to be known or what could be known about the situation, to see whether it might be interpreted in any coherent way. The narration includes summarizing what is known about the machine, questioning whether what is known is a coherent representation of the situation or could be if interpreted differently, and determining what else needs to be known and how to learn it. The question of interpretation is critical; technicians working on incomplete diagnoses often say they believe that they know the crucial facts but do not recognize them. The hard part is recognizing the significance of any given fact with reference to the observed or reported behavior of the machine. The narrative includes a formulation of how the described state of the machine produces the behavior described by the customer and may include a hypothetical history of the machine and its use which would lead to that state.

This narrative, piecemeal production of a representation is due partly to the nature of the machine and partly to the nature of information about the machine. Copiers are complex, elaborate assemblages of mostly simple mechanical or electromechanical components whose functioning may easily be tested. Many of these tests require a kinesthetic awareness of the machine, knowing whether the feel of a given mechanism is right or not, but this does not appear to be a problem for the technicians observed. Those components that are not simple are either treated as black boxes or treated as the sum of their observable functions and are thus rendered simple. Some problems are solved in testing the simple components. These relatively simple components are, however, strung together into long, complex, interactive chains; diagnosing such a chain requires both the patience to follow it out, testing each piece in turn, and the ability to maintain control of the results and of the cumulative significance of such tests. For harder problems, the difficulty in representing such a system lies not in knowing whether a given component or subsystem is working but rather in the ordering and integration of one's knowledge of many different components or subsystems into a meaningful representation. This includes remembering how many different subsystems influence a given process and knowing whether all have been considered. Thus, the information available from the physical state of the machine consists of relatively simple facts, which if not known to be critical by themselves, need to be integrated with other such facts in order to be significant.

The technicians have many resources for diagnosis—the customers, the state of the machine, the logs, the documentation, their experience and that of their colleagues—but none consistently supplies definitive answers for all problems. It may not be possible to find anything wrong with the machine, and yet the machine will not run. The customer may not be able to say what was happening, but the machine did not work as expected. The problem may be still worse: the documentation may be incorrect or may simply omit some information. Errors recorded by the machine may reflect problems other than those they are intended to report, or what is happening in the machine may not create a record at all. Entries in the log are as reliable as one's cohorts, but the log is a public record in which a technician may want to show that which was supposed to be done, whether it was actually done or not. No one may ever have heard of any problem quite like this one. For perhaps the majority of service problems, the various diagnostic resources supply an answer, providing an adequate representation of the situation so that the technician may fix it. For the truly difficult problems, they supply bits of information which are raw material for the technician to use in creating a representation.

When diagnosis requires production of a full representation of the situation, this production appears, in fact, to be a form of *bricolage*. Lévi-Strauss writes of the *bricoleur*:

> His universe of instruments is closed and the rules of his game are always to make do with "whatever is at hand," that is to say with a set of tools and materials which is always finite and is also heterogeneous because what it contains bears no relation to the current project, or indeed to any particular project, but is the contingent result of all the occasions there have been to renew or enrich the stock. . . . Consider him at work and excited by his project. His first practical step is retrospective. He has to turn back to an already existent set made up of tools and materials, to consider or reconsider what it contains and, finally and above all, to engage in a sort of dialogue with it and, before choosing between them, to index the possible answers which the whole set can offer to his problem. (Lévi-Strauss 1966, pp. 17–19)

Lévi-Strauss's intent was to use practice as a metaphor to describe modes of thought, but the result is an apt description of a way in which work gets done. The point of *bricolage* is the reflective manipulation of a closed set of resources to accomplish some purpose. The set is the accumulation of previous manipulations, one's experience and knowledge and, in literal *bricolage*, physical objects. This manipulation is done in the context of a specific goal, which influences the process. The items in the set are not limited to a single use or a single meaning, but their properties limit their possible applications. The point of the manipulation is to see whether a given item from the set may be applied toward the goal of the *bricolage*, possibly by some reinterpretation or modification, or whether it has some quality that makes it inappropriate in the present context. For my purposes, the significant aspect of *bricolage* is the reflective manipulation of a set of resources accumulated through experience, with the range of manipulation neither totally free nor constrained to the original manifestation of any element. Like the *bricoleur*, the technician has a closed set of information resources that do not necessarily provide definitive answers. The bits of the puzzle must be examined in the light of experience to see which combination provides the most reasonable representation of the problematic situation.

The majority of the problems encountered in the service calls that I studied were solved quickly, with a few observations yielding an obvious culprit. Some were well known, as shown in the earlier transcript; after conversation with the customers provoked suspicion, opening a cover to feel the free play in a shaft confirmed the diagnosis. Other problems were inescapable, such as mechanical obstructions completely preventing operations of the mechanism, or damaged components immediately obvious when the mechanism was inspected. In still other instances, the reported error codes were accurate, and the problems associated with them in the documentation were the true culprits. Some of these were diagnosed with the aid of the documentation, others from the technician's memory. In one case it was known that the documented cause was only one of those that could produce the symptom, but the others were much harder to find. The technician taking the call was very relieved when the official cause turned out to be correct.

Other problems were dissolved, that is, made to go away without specific solution. The technicians knew that a series of routine maintenance procedures would eliminate almost all problems of a certain class and were also the standard preliminary procedures in correcting any of them once diagnosed. By doing the preliminary procedures before the diagnosis, they eliminated most problems and were better prepared to do more serious diagnosis on any that should remain, but during the period of observation there were no problems left at the completion of the dissolution procedures. These problems are not solved, in that one never knows what the particular problem was, but they are gone, dissolved by the technicians' actions.[1]

Dissolution was attempted in other instances, either when diagnosis failed to produce a cause or when the machine refused to display the problem and the customer's formulation of the problem was less than definitive. These attempts were less successful. In such circumstances, the range of potential problems and causes is too great to test any significant percentage. The dissolution strategy does eliminate some of the usual suspects should the problem recur, and it may be lucky and successful. Perhaps more important, the customer has reported a problem, and it is the customer's problem which the technician must address, whether the machine is obviously broken or not. The technicians feel that they need to be seen to have done something as part of their contract to address their customers' problems with their machines.

A failed diagnosis is a failure to construct an adequate account of the problem. As suggested above, the nature of causality in the machine is such that crucial information may be either unavailable or not of obvious significance in a given situation. The technicians have models for some machine problems with indeterminate causes, associating them with some known possible fixes. These fixes are as much dissolution strategies as solutions, given that their association with the problem is partial and tentative. They are known as one solution to a set of problems that produce more or less the same symptoms. Unfortunately, when the fixes have been tried without improving the situation, and doing so again yields no better results, no further means to improving one's understanding of the problem are readily available. The usual recourse is to bring in another technician or technical specialist on the premise that a fresh perspective may make it possible to reinterpret the known facts into an adequate representation of

[1] The concept of dissolving problems arose in conversation with Suchman, but derives from the work of Jean Lave. Lave's study (1988) of the arithmetic processes of grocery shoppers suggests that shoppers employ a series of gap-closing procedures which transform problems into ones that can be resolved, rather than seeing their problems in terms of pure arithmetic. In these terms, dissolution transforms problems out of existence, rather than into a form susceptible of resolution.

the problem. However, there is no guarantee that this will be successful, and the possibility of real diagnostic failure is always there.

Perspective is important in diagnosis. As I have suggested earlier, with these machines the problem is not so much testing as maintaining control of the results of those tests and interpreting them. In most of the hard diagnoses I observed, solution was discovered through reinterpretation of known facts and following the new interpretation with new investigations. This is one of the reasons that consultations and joint troubleshooting are so popular and effective. The presence of someone else guarantees another perspective and makes it easier to experiment with new interpretations. It also provides someone to whom stories can be told and who will tell stories in return; the telling of war stories, the consideration of the present with reference to known diagnoses of the past, is an essential part of diagnosis.

8

War Stories of
the Service Triangle

The use of war stories is a prominent feature of diagnosis among the technicians. These stories are anecdotes of experience, told with as much context and technical detail as seems appropriate to the situation of their telling. At a minimum they name the technician doing the work, the machine to which it was done, the problem, and its solution; in the majority of cases I observed, the technician telling the story is the one to whom it happened. The stories occur naturally in discourse among the technicians, either in diagnosis or in more purely social situations. There seems to be little question of gaining the right to tell stories (Sacks 1970, 1972, 1974); discourse among the technicians appears to presume that all competent members of the community will tell stories.

Telling stories in diagnostic contexts makes some of them extremely elliptical and barely recognizable to outsiders as stories. The ellipsis is permissible because the context will be used in interpretation to supply some of the missing detail, as will the common experience of teller and listeners, and because, in an interactive situation, the teller can count on the hearers to indicate if the ellipsis is too great. Such excessive ellipsis is easily corrected through normal conversational repair. This brevity is a matter of cultural propriety and competent practice; it would be inappropriate to waste everyone's time repeating the superfluous to make a well-structured story, particularly in the context of a service call when the technicians' sense of their own professionalism requires a speedy resolution to the situation. It is also true that these elliptical stories provide all the essentials for those sufficiently versed in the world of service to fill in the rest.

Once war stories have been told, the stories are artifacts to circulate and preserve. Through them, experience becomes reproducible and reusable. At the same time, each retelling is, in a sense, a re-representation. The stories originate in problematic situations and are told or retold in diagnosis when the activity they represent becomes problematic again. They are retold in the consideration of a present problem, when the issue of comparability of context with some previous experience has arisen, and this renders the previous, completed episode once more problematic.

War stories are also told in pursuit of more purely social functions than diagnosis. They preserve and circulate hard-won information and are used to make claims of membership or seniority within the community. They also amuse, instruct, and celebrate the tellers' identity as technicians. Indeed, they are offered in response to questions from inquisitive ethnographers, whose questions may make the most mundane activity problematic. In more normal social discourse, the problematic quality that occasions the telling of the stories seems to stem primarily from a wide range of occasions on which technicians are called upon to account for their activities and secondarily from a need to represent themselves in a heroic or at least competent perspective. In these tellings, past problematic circumstances are made publicly and collaboratively inspectable by one's peers, and one's experience is made reproducible and reusable on subsequent occasions by others. Such tellings are also demonstrations of one's competence as a technician and therefore one's membership in the community.

War stories are told in diagnosis when no clear formulation of the problem is emerging from the welter of facts. Technicians may find with some problems that they know a great many things about the machine but that the facts do not add up to a clear picture of the problem. Telling stories of more or less similar experiences is a way of pushing the facts around, trying other perspectives to see if they suggest other interpretations. As discussed in the previous chapter, the available information about the machine is not necessarily reliable and frequently less than clear; war stories have a great advantage in credibility because they recount the personal experience of a specific individual, whose competence is known and can be considered along with the material about the machine. Such stories combine facts about the machine with the context of specific situations. The contextual information demonstrates the claimed validity of the facts of the story and provides a framework for testing those claims against the hearer's model of the machine. The context also constrains the application of those claims by defining the situation in which the facts are known to be true. War stories told during difficult diagnoses are doubly situated, first in the context of their origin and then in that of their telling and possible application, and the comparison of the two situations is the point of the telling.

There is an inherent uncertainty in the situation of diagnosis. In difficult diagnoses the technicians are contending with the limits of their knowledge. The known weaknesses of their sources of information suggest that they may be presumed to have an incomplete set of facts about the machine; they lack the perspective of understanding which would integrate those facts into a coherent representation, which might or might not indicate the need for additional facts. This is a double-edged coherence, requiring both coherence of story and connectivity of the facts. Without coherence, the technicians cannot know whether they have all the facts they need. They do not know whether they have overlooked something, whether there are more facts to gather, or whether this is a new problem that is completely beyond their experience and understanding. They do not know whether they lack a fact or an interpretation. Their experience suggests the latter, that the answer lies in gaining the proper perspective on things they already know; in pursuit of this, an amazingly wide range of stories may be told. Some appear to be startlingly irrelevant until one realizes that the point is not just to consider the symptoms but also to jar one's perspective into focus.

Some of the stories told during diagnosis are clearly used to eliminate suspects. There are conflicts between the observed facts about the machine and those informing such a story. It would appear that these stories are told to spell out the differences and to stress that the present situation does not fit the other, known problem, so one may quit thinking about it. Such stories may also have a wistful quality, as if to say that the teller wishes the problem were that simple.

Other stories are primarily exhortations to think clearly, warnings that failure to remember the sometimes invisible or illogical connections between symptoms and causes may add hours of unnecessary diagnostic activity. One such story was told during a joint troubleshooting session, a story about the importance of paying attention to details, telling how failure to recognize the significance of a minor machine noise could mean long hours of diagnosis. This story was told as part of an inquiry about the presence of that same easily overlooked machine noise, serving both as text to emphasize its importance and as recognition of the ease with which it can be missed. The two technicians present had two different versions of the story, revealing interesting variations in the way technicians understand machines, so I shall include the transcript and discuss the episode in some detail.

The beginning of the story by one technician provoked a show of recognition from the other, and the story unfolded in antiphonal recitation. I consider these different tellings to be the same story much as different versions of a folktale are considered to be the same story. The problem is the same, as is the path to solution, so the plot is constant. The protagonists are categorically identical, a machine and a technician; changing the name of the technician does not really change the story any more than changing the name of Cinderella does. The technicians themselves focus on the fact that the problem is the same, not that different people have experienced it.

The two technicians were working on a machine which had recently been installed in a new building; it had never worked reliably. The recurrent failures produced a specific type of error message; however, changing the components indicated by the documentation did not change the behavior of the machine.[1] This particular customer had many machines and even an assigned

[1] The documentation has diagnostic procedures associated with each error message. Following these procedures had led the technicians to change components but had not solved the machine's problem.

technician, making the potential cost of failure to solve this problem greater than normal. The team's technical specialist had joined the technician responsible for the machine in addressing the latest appearance of this problem.

There are some technical terms that appear in the transcript, and it may be useful to differentiate the categories of interest. First, there are components; those mentioned are dicorotrons (or dicors, for short), the shield, which is part of the dicorotron, and the XER board, one of the machine's circuit boards. The 24-Volt Interlock Power Supply is not really a component but is the term for an output from a power supply controlled by numerous safety interlocks. The observation that the relay switching it has opened is significant because it usually indicates the opening of one of the interlocks. Next, there are error codes, such as E053 and F066; these are diagnostic error codes, and the documentation contains procedures to track down the causes of each code. Unfortunately, some error codes can be produced by problems other than those anticipated by the documentation. Finally, there is the dC20 error log, a record maintained by the machine of various problematic events in its history. The accumulation of entries in this log is very useful in diagnosis. The telling of the two versions of the story was triggered by the observation that the characteristic failure of the machine being diagnosed includes the opening of the 24-Volt Interlock Relay.

Technical Specialist: See, this runs along with the problems we've run into when you have a dead shorted dicorotron. It blows the circuit breaker and you get a 24-Volt Interlock problem. And you can chase that thing forever, and you will NEVER, NEVER find out what that is.

Technician: Yes, I know, E053, try four new dicors. . . .

Technical Specialist: But, if you went in . . . OK, you won't . . . You lose your 24, that's what it is: you're losing your 24-Volt out of the power supply, but that's not what it's caused by. Now the key there, though, is when you pull up your dC20 log, you get hits in the XER board.

Technician: Yeah. The other thing is as you're going on and on and getting E053s, you get, yeah . . . F066 . . . in the sequence. . . .

Technical Specialist: If you're lucky enough for it to run long enough, you'll get an F066 problem which leads you back into the dicorotrons—you check them—yeah, I've got one that's a dead short. You change it and everything's fine, but if you don't . . . if you're not lucky enough to get that F066 or don't look at the dC20 log, it's really a gray area. . . .

Technician: Well, dC20 logs . . . when I ran into that I had hits in the XER a few times previously, so I was tending to ignore it until I was cascading through after an E053 which is primary, I'm cascading to see what else I've got—F066—what the hell's this? Noise?

Technical Specialist: E053, which one's that?

Technician: Well, that's a . . . that's a 24 . . . lock

Technical Specialist: 24 Interlock failure? Yeah. We did . . . I did that not knowing when they changed the circuitry in the XER board, normally if you had a shorted dicorotron, it'd fry the XER board—just cook it. Now they've changed the circuitry to prevent frying of that, but now it creates a different problem.

Ethnographer: This is with your dicor shorted to ground or . . . ?

Technical Specialist: Probably shorted to DC shield.

Ethnographer: Ah hah, yeah. . . .

Technician: Mm. I see.

Technical Specialist: OK, and then that goes [*snaps his fingers*], you know. That's where it's popping the breaker, and when it does that, that's when you end up with . . . through the boards, it pops it, before it pops the breaker, because you don't have any DC boards . . . you'll get a DC Interlock, . . . 24-Volt DC Interlock failure. Now that came about after these boards came out and I've gotten burned twice on that same problem. I guess I . . . that four hours of sticking my head in the machine and tracing 24-Volt Interlock problem the first time didn't do it. The second time, it took me a long time, and it finally dawned on me—what the hell am I doing—get in there and that's what it was. Now I've had very intermittent dicorotron problem— same thing—guy sits there, and he says "I've got . . . ," he says it's intermittent, once, maybe twice a day; you shut it off, turn it on, and it will run . . . just a 24-Volt Interlock failure problem. So I asked him about the dC20 log and he said "Yeah, I had

XER failure"—I says OK—he says, "I checked all the dicorotrons"—I said, you're going to have to stress test 'em for a long
period—four to five minutes in the high . . . you know, in your
dicorotron checkout where you really, you've got the currents
boosted up on 'em . . . and yeah, after about four minutes one
of them [*snaps his fingers*] glitched.

The gist of the story, then, is that there is first a red herring, an
E053 error code which is not to be believed, which may be followed by a second error code, F066, indicating the true culprit, a
shorted dicorotron. Failure to remember the connection between
the deceptive E053 and a shorted dicorotron could mean a long,
frustrating, futile attempt to diagnose the problem. This is an example of telling a story as a reminder of the tenuous connection
between some symptoms and their causes. The technicians are
faced with a failing machine displaying diagnostic information
which has previously proved worthless and in which no one has
any particular confidence this time. They are looking for some
link from the error message to the real problem, whatever it may
be, and this story reminds them of the elusive nature of such
links.

The progress of the responses in this dual recitation under-
scores the point that storytelling must be seen as an interaction
between teller and hearer (Sacks 1974; Smith 1980). Both versions
are based on personal experience but reflect significantly different
views of the problem and apparently different approaches to the
task of understanding the machine. The technician's version tells
of a pattern-matching diagnosis, which associates a solution with
the first error code: "Yes, I know, E053, try four new dicors. . . ."
The appearance of the second but more important error code is
not seen as problematic: "The other thing is as you're going on
and on and getting E053s, you get, yeah . . . F066 . . . in the
sequence. . . ."

In the technical specialist's version, the appearance of the second error code is very doubtful: "If you're lucky enough for it to
run long enough, you'll get an F066 problem which leads you
back into the dicorotrons." For him, the problem is more reliably

indicated by apparently random and insignificant entries in a particular part of the error log: "Now the key there, though, is when you pull up your dC20 log, you get hits [*logged events*] in the XER board."[2] He suggests that even testing for the true fault may be problematic, requiring longer than usual periods of testing to provoke the fault under some circumstances: "He says, 'I checked all the dicorotrons'—I said, you're going to have to stress test 'em for a long period—four to five minutes in the high . . . you know, in your dicorotron checkout where you really, you've got the currents boosted up on 'em." His version includes both the evolution of the problem, that it is the result of a fix preventing the same fault from destroying parts of the control system, and evolution of the understanding of the problem through his own history of performing the diagnosis.

The technical specialist's version associates the problem with a system, the 24-Volt Interlock, while the technician associates it with an error code, and this difference seems significant. The technical specialist is thinking about the 24-Volt Interlock system and uses that as the entry point for this story about the same system. He does not make the connection to the error code E053, but then it is not obviously relevant because that is not the error code facing them. The technician recognizes the story, perhaps by the reference to a shorted dicorotron, but his pattern-matching version is based on the error codes. The link to the present situation is less obvious because the link of the error code to the 24-Volt Interlock system is hidden, and he stumbles when asked what the error code indicates.

There is a curious alternation in voice to be observed in the telling of the two versions of the story. Both tellers initially favor a generic second-person presentation: "See, this runs along with the problems we've run into when you have a dead shorted dicorotron." In this utterance one finds both a "we" for the community experience and a "you" for the abstract technician encountering the problem. At this point the story concerns a class of problems that occur on this class of machine, and there is no direct reference to personal experience. ("I know" in the technician's first utterance seems to express recognition of the problem

---

[2] The XER board is both a component and a category in the dC20 log.

more than a claim of personal experience.) The two continue with this impersonal second person until the technical specialist says: "If you're not lucky enough to get that F066 or don't look at the dC20 log, it's really a gray area."

This prompts the technician to change voice and tell his personal experience: "Well, dC20 logs . . . when I ran into that I had hits in the XER a few times previously, so I was tending to ignore it until I was cascading through after an E053 which is primary, I'm cascading to see what else I've got—F066—what the hell's this? Noise?" He had been sufficiently lucky to have the secondary error code which indicates the true problem. The last long utterance by the technical specialist begins in the second person and switches to the first to tell about the difficulty of remembering the association, a point he had been making earlier and probably the point of this particular story. Thus, the second person is used here to describe a specific problem in a somewhat abstract way, while the first person is used for personal experiences. This alternation of voice is also a movement from the less situated second person in no particular context to a situated first-person account, and it seems to be the awareness of situated experience that gives credibility to the more general accounts.

This is a fine example of the ellipsis discussed earlier; the story is so elliptical as to be only marginally recognizable as a story. It is clear that the technical specialist tells of two encounters with the problem, both successful, but there is nothing to indicate where or when they occurred, or how the first encounter was solved. Solution to the second occurred through the grace of a reluctant memory reminding the technical specialist that this had happened before. The additional detail about the origin of the problem and its known variant is also typical of technicians' stories when used in diagnosis; detail is preserved and told because one can never be sure which piece will complete the next puzzle.

The principal theme of the technical specialist's version in this telling is that the problem as perceived is insoluble because the error code is wrong. This also describes the problem confronting them. The lack of logical connection between the error code and the real problem makes the connection extremely fragile, as attested in the last utterance here, describing a second period of troubleshooting the false problem until memory intervened. This recognition of tenuous and arbitrarily recovered connections between symptom and problem may serve either as inspiration or as a warning in their search for the real problem, whose link to the present false symptom is either forgotten or undiscovered.

At other times, wishful thinking leads the technicians to tell stories about causes they would like to find. Later in the same troubleshooting session, the apparently random distribution of a particular type of error message led the technicians first to wonder about the AC supply to the machine and then to dismiss the idea because problems there would be unlikely to produce such consistent error messages. Eventually, lack of any other suspect led the technicians back to the subject of the AC supply; in thinking about it, they told half a dozen stories of various AC problems. Apart from their disqualifying caveat, there were several valid reasons to be considering the AC supply at this point: first, AC problems often occur in new buildings like this; second, these particular machines have a history of AC problems even in older buildings because of internal differences in the way they switch the AC; third, the technicians were running out of known possible causes and were seeking to expand their problem space; and fourth, if the AC were the problem, it would be someone else's responsibility. With reference to the last point, several of the tales also featured the recalcitrance of electricians and their reluctance to accept that there might be problems with the AC or its distribution or that these problems might cause difficulties for the copiers in the building. This series of stories went on until it was apparent that they offered no inspiration and that there was no reason to ignore the earlier disqualification. At this point it was deemed time for lunch, in hopes that something different would happen to the machine in the interim. When we returned from lunch, *mirabile dictu,* there was a new error code that pointed us in the right direction.

A final category of stories told in diagnosis are those that support a developing diagnostic understanding. That is, the technician has thought of a new gloss for the known facts; the question is whether it makes sense. As actually uttered, the support may be less a story than an affirmation, sometimes in detail, that one has encountered such a combination of cause and effect before; however, I believe that a better developed story could be elicited. Such stories also appear during consultations. Should a recommended course of action be thought improbable, the recommendation may be augmented by short accounts of how it worked for some other known technician. Similarly, if a suggestion of a possible cause for the problem is met with the protestation that that is simply impossible, the response is likely to be a story in which the problem is exactly that, despite the same sets of objections.

War stories are also told extensively in contexts other than diagnosis, where they may be seen as representations of one's understanding of the world as it is or perhaps as it should be. Some of these are merely illustrative, explaining to the ethnographer or another neophyte how one manages a difficult customer. One story was a claim to have survived the initiation hazards and become a full-fledged technician for a given machine. What is most interesting about this story is that it is a celebration of conspicuous failure, which is certainly as much a part of the technician's fate as unrecognized success:

> Until you really break something good, you're not a member of the team. I became a member of the team the day I smoked a transformer because they wired the, uh, cord at Aerospace . . . they put different caps on the cords. They [*manufacturing*] wired it neutral on a hot line, and I put in the machine with the neutral on a hot line, and a hot on the neutral. . . . And I installed it, dutifully, and didn't know there was a, uh, an actual program for installing the cord. And I didn't trust them, and checked the wall, but it did not occur to me to check the terminals, right, and that of course is something that I'll never do again. You should have seen it: I had my back to the machine and smoke is rolling out of the card cage. And the lady says, "OHHH, I think we have a fire." And I say, "Oh, no problem," and went over and unplug it and said, "OK. Now I am a member of the team." All it took out was, probably all of the things in the Low Voltage Power Supply that had to be replaced, probably only that one transformer was gone.

Other stories are told as a challenge, to see if the other technician is good enough to recognize the situation described. The result of this may be disconcerting. In one instance, the person to whom such a story was told did indeed recognize the situation, but the person telling the story had it from someone else and had not yet decided whether it made sense. There are still other stories that clearly express a moral attitude about how one should deal with a service call. They tell of finding technicians in positions they should never be in, with threats not to help in the future until the technicians in question have been stuck suffering for several days.

Stories are also told to instruct. I heard the False E053 story told this way three months after the antiphonal recitation above. This happened during lunch at the branch office; the technical specialist encountered in the previous dialogue was playing cribbage and talking with various passersby. One of the technicians on his team, not the same technician as in the original episode, was going to have another researcher observing him the following day, and asked:

Technician: It'd be nice if I get an E053, . . . put in a short somewhere in the machine, that'd be good troubleshooting. Sam [*the technical specialist*], why don't you go to HighTech [*a customer site*], Friday evening, and put in an E053 code, so I could find it, show this guy some troubleshooting.

Technical Specialist: You don't want me to do that.

Technician: I know, 'cause you'd probably make sure I couldn't find the damn thing and ruin my whole week.

Anonymous other technician: A short somewhere is kind of vague.

Technical Specialist: No, it's not, not when you understand wiring.

Ethnographer: Oh. Well, there was that one that we had [*the first technician, when I was observing him*], over at that first place you and I went, where there was the short because the RDH harness was in the wrong. . . .

Technician: Right.

Ethnographer: That's what you had in mind?

Technical Specialist: No [*pause, deals*]. The common one is when you have a shorted dicorotron. . . . Can't even get the machine to run, it'll cycle about twice and then shut down, and give you an E053.

Ethnographer: Oh, well, that's too easy. [*Pattern-matching . . .*]

Technical Specialist: Well, it wasn't the first time I had that. [*. . . when the pattern is not yet known.*]

Ethnographer: Where does the dicorotron short, to the shield?

Technical Specialist: Direct arc [*pause for cribbage*].

Ethnographer: From foreign matter or just, ah . . .

Technical Specialist: . . . bad glass on the wire or something, a direct arc. . . . [*but he doesn't say where to; long pause for cribbage*]. First time with the new boards, the new XER board configuration, it wouldn't cook the board if you had an arcing dicorotron. Instead, now it trips the 24-Volt Interlock in the Low Voltage Power Supply, and when it comes back . . . the machine will crash and when it comes back up it'll give you an E053. It may or may not give you an F066 that tells you the short is in, you know, check the xerographics. That's exactly what I had down here, at the end of the hall, and Weber and I ran for four hours trying to chase that thing. All it was was a bad dicorotron. We finally got it, . . . run it long enough so that we got an E053 with an F066, and the minute we checked the dicorotrons we had one that was totally dead. Put a new dicorotron in it and it ran fine [*long pause for cribbage*]. Yeah, that was a fun one. That's . . . the first time . . . you know usually, if you have a real bad dicorotron, you used to cook the boards, generally more than one. So, saved us buying a lot of boards.

It is clear from the start of this exchange that both the technician and the technical specialist know they are discussing the same problem, and the association of error codes with the problem is accepted by both. The story is now much more concise and didactic, and the difference between this version and the previous two may be due to the situation of its telling. To begin, the scene is not one of crisis but occurs during a time of relaxation, backstage at the branch away from customers and other noninitiates (with the partial exception of the ethnographer), nor is there any particular machine involved. In this context, the technician feels free to express his desire to demonstrate spectacular troubleshooting prowess while being observed by an outsider. It should be noted that such a display is only possible for an outsider; to another technician he would merely be doing a competent job. However, given the inherently unpredictable nature of technicians' work, nothing interesting may happen, and a planted problem could liven things up considerably for the visitor.

Such planted problems are used in school as part of the technicians' training. However, the technicians are fond of the saying, "If it ain't broke, don't fix it," and this wisdom certainly extends to not breaking those things not yet broken. Their agreement on this recognizes both that there may be unforeseen consequences to perturbing a working machine and that the social situation is not that of school, where such consequences could be tolerated and used for learning. In the working environment, the machines are placed with customers for the customers' purposes; the social contract with the customer is based on the customers' need for outside expertise to keep the machine running and the technicians' possession of that expertise. The machines are sufficiently complex, and causality sufficiently vague, that one cannot be entirely sure of the consequences of doing things to them; even repair actions may change machine behavior in unpredictable ways. This is an entirely acceptable risk when a technician is repairing a broken machine, but given these considerations, planting a shorted dicorotron in a working machine to produce the False E053 problem would be foolish.

The story does not appear in this preliminary exchange between the technician and the technical specialist because both know whereof they speak. The story is triggered by the fact that both an unidentified other technician and the ethnographer reveal their failure to understand the significance of the short that causes E053s. Part of the technical specialist's job is to see that the technicians keep their skill and information current, and he wanted me to understand what was going on, so the technical specialist responds with this retelling of the False E053 story. This version should perhaps be viewed as a description of how the community ought to understand the problem.

The clear point of this telling is to emphasize to the specialist's listeners the importance of associating false E053 error codes with shorted dicorotrons; the use of the first person with so much situated detail—who else was with him and which specific machine it was—lends authenticity to the rest of the account. Some of the detail about the origin of the problem (and the disappearance of the cooked board problem) may appear extraneous, but it does enrich the hearer's understanding of the machine, and one never knows which piece of information will prove crucial as new problems continually appear. The apparent object of including peripheral detail is to keep all knowledge as closely connected as possible, so that if a new problem connects to any known facts at all, it connects to an understanding of the system, with known failures and solutions on which to base a diagnostic strategy.

There is another class of war story that can only be seen as a celebration of being a technician, able to cope with anything that either machines or customers or both can do. Some of these have pedagogic connotations as well. One such story describes how a customer used the wrong supplies in a machine and chose ones chemically incompatible with the machine. The problem was solved by a fortuitous encounter with the customer who had done this, and the story celebrates this triumph, while warning of the unimaginable depths of customer failure to understand how to deal properly with the machine. Other stories, especially those of the past, are clearly more celebratory. These tales of the heroism required to service early machines seem balanced between celebrations of the perversity of the machines and celebrations of the technicians for coping. It is not clear whether the technicians more admire the coping or the perversity.

Such stories are also told to challenge the new technicians, offering examples of past heroism with the clear implication that these opportunities are no longer available. Tales to new technicians, however, are more than claims to superior status or attempts to intimidate. A war story is both a model of a service call and a model for a service call, in Geertz's terms (1973). It tells them how a call should evolve, in ways that they can use during a call, influencing the call to evolve in that way. As such, war stories may be a valuable part of the transformation of a new hire into a technician.

As mentioned earlier, one hallmark of war stories is their elliptical style. Every service call is roughly the same, differing primarily in details. Since the stories are normally told only within the community of technicians, there is no reason to retell any of the structure of the call; that is presumed by one's listeners. What remains is what the community regards as storyable, remarkable (Sacks 1970); what the technicians tell is what they are entitled to bring away from the situation, and learning this is also part of one's initiation into the community. The details are what interest the technicians; they have become connoisseurs of the variations in machine misbehavior and of new shades of misunderstanding displayed and practiced by the customers. Accordingly, most stories set the scene by naming the machine, which conveys an entire social situation to those who know, and then plunge into the details of the problem. The only other point that matters is that the technicians cope and that this triumph be known. Since every service call is at least somewhat problematic, stories of how one of their number solved yet another puzzle are important in building the confidence that they, too, can solve the next problem.

Almost everyone in the corporation knows that technicians tell war stories. The official attitude toward war stories varies directly with distance from the field. As I mentioned earlier, technicians told me that their immediate managers delayed starting team meetings because the technicians were telling each other stories about their most recent experiences, and this information exchange was perceived as useful or even vital. Those who have been promoted away from the field tend to see the storytelling as purely social and political, making claims to membership or position based on experience; the stories that exchange information are perceived as categorically different from war stories. With some notable exceptions, those who have never worked in the field but know of the notorious war stories tend to perceive them as a way of killing time when one should be working.

The ways in which technicians tell and use stories are not unique to field service; the literature on stories reveals some interesting comparable examples. Just as technicians tell stories to make sense of the diagnosis confronting them, to bring their prior experience to bear on their current problem, and to preserve what they have learned, Evelyn Early (1982) finds that traditional residents of Cairo use narratives to set illness in context, to evaluate options of treatment, and generally to make sense of their situation. The technicians shape their narratives to create meaning in an inherently ambiguous situation full of facts that do not obviously make sense; the courtroom analysis of W. Lance Bennett and Martha Feldman (1981) suggests that trials be seen as competing narratives interpreting approximately the same set of facts but making different kinds of sense. The telling of stories is a situated practice, and some of the differences in the versions of the False E053 story stem from the different contexts of its telling. Brigitte Jordan (1987) found that Mayan midwives' stories about birth were normally told while attending a birth, as part of the process of deciding what to do next. Outside of the birthing context, such stories were only told if they came up through some other contextual tie, such as passing the house of one of the participants, or through relationship of one of the principals to someone else with whom the midwife was working. The technicians were never seen apart from the context of service, so it is not known whether stories would then be difficult to elicit, but the immediacy of context obviously affects their stories.

Technicians clearly tailor their stories to the occasion of their telling, being as concise or as elaborate as necessary for their hearers or their needs. Stories that work well for members of a culture may seem remarkably cryptic and elliptical to outsiders. Renato Rosaldo (1986) describes the basic Ilongot hunting story as a sequence that names places and activities, presumably in the chronological sequence in which they were visited or done. Common additions include the preparations to hunt, the success, or the weather. Ilongot feel that stories require specific geographic references to passes, mountains, and river crossings, but given those, any competent listener would know just what is involved in traveling over that terrain and doing the activities named. Rosaldo finds these stories particularly revealing of Ilongot culture in that they emphasize the portions of hunting experience that Ilongot find interesting, omit those portions that every culturally competent listener knows, and finally celebrate the hunter's prowess. While the technicians do prune their stories to omit what they consider obvious, they preserve a startling amount of detail, suggesting exactly how much of each episode is problematic and possibly essential to the contingent truth they intend to tell.

Rosaldo also suggests that part of the attractions of the hunt are the paired opportunities first to display the taut alertness and quick improvisation so highly valued by Ilongot hunters and then to recount this in a tale. This is a legitimate opportunity to portray oneself as a star in one's own culture. In their discussion of occupational communities, John van Maanen and Stephen Barley (1984) describe a similar mechanism to achieve status by resourceful improvisation, which is then material for stories told to the rest of the community. Technicians clearly tell tales to establish their membership in the community, and the tales which preserve their solutions to problems also contribute to their reputation. Their individual work would not be known if they did not tell it; other technicians would assume that the service call was routine. However, the stories rarely seem to celebrate the individual as star; they seem more focused on the heroism of mere survival and thus are more involved with the dimensions of the machine's or customer's misbehavior. The emphasis is on the perversity of the world in which the technicians work, and while each instance is usually resolved by an individual, celebration seems to be for the community.

This celebration of the technicians' competent and collectively heroic practice is, perhaps, necessary as a counter to the image of service presented by the corporation in their insistence that service work is merely following directions. A comparable use of story to insist on one's identity is reported by Barbara Myerhoff (1986), who recounts the actions of a community of elderly Jews, survivors of the Holocaust, who felt themselves losing identity, dissolving into the scenery of a society that did not know they were there. Their presentations of their lives and liveliness effectively halted the dissolution, perhaps because they had convinced themselves that they lived. "One of the most persistent but elusive ways that people make sense of themselves is to show themselves to themselves . . . by telling themselves stories. . . . More than merely self-recognition, self-definition is made possible by means of such showings, for their content may state not only what people think they are but what they should have been or may yet be" (Myerhoff 1986, p. 261). It seems clear that one motivation for the technicians' stories is to present themselves in a mode that they like. The image of service work as presented from outside their occupational community is one they disown and dislike. Their own presentations of their skilled work allow them to claim their identities as practitioners of the literally black art of restoring harmony to the relationship of customer and machine.

These stories are part of the occupational community (van Maanen and Barley 1984); they have little to do with the corporation as a whole. In contrast to Joanne Martin's work (1982) with stories told in support of an organizational ethos, the organization rarely appears in technicians' stories, but then the organization is largely irrelevant to the technicians' actual work, which is performed alone or with one or two companions. This promotes a gunslinger mystique of self-reliance: the lone technician walks into the customer site to cope with whatever troubles lie therein . . . but with the community available as a resource. The technicians are both a community and a collection of individuals, and their stories celebrate their individual acts, their work, and their individual and collective identities.

However, it is crucial to note that stories do more than celebrate the job; they are part of the job. In both Early's work (1982) with stories told to resolve health-care decisions and Jordan's work (1987) with the role of stories in childbirth, the narratives are part of people's practice, a role which has not been acknowledged for them in industrial work. Stories in occupational communities are said to be about work but in the interests of membership or stardom, while Martin's stories are about the organization. Technician's stories *are* work; they are part of diagnosis, and they help preserve the knowledge acquired for the benefit of the community. Stories are more than a celebration of practice; they are an essential part of the practice to be celebrated.

9

Warranted and
Other Conclusions

Community Values

To conclude this discussion of technicians' practice, we must consider the values reflected in their discourse and their stories. Given that two principal issues for technicians are the fragility of understanding and the fragility of control, it is not surprising that many of their war stories and most of their values hang on those issues. Each war story of a problem solved is another situation preserved from chaos, which should be celebrated, while simultaneously reminding teller and listeners that the peril is still there. Technicians' discourse reveals that they value most highly those attributes that contribute to the preservation of order and understanding. Reputations are built on technical skills, memory, ability to gather information, verbal performance, and the general ability to retain control of the situation.

This concern for control is paralleled by a drive both to preserve order in one's work and to appear always to be in control of the situation. We have seen that a technician who walks up to the wrong machine at a customer site cannot walk away but must do something, perform a minor service and checkup to maintain the appearance of control. Similarly, it seems important to technicians to keep tools and parts in order, to keep track of them and not have them scattered all around the room. It is not clear whether the reality or the appearance of order matters more, nor whether

144

it is important for customers, colleagues, or themselves, but order is clearly valued as a hedge against chaos. The technicians have an almost Sisyphean view of the probability of chaos, as shown in this quote: "One week everything's OK; the next you have calls up on all machines, all of them call-backs. . . ." Given this concern for order, it is not surprising that technicians who are having difficulty are legendarily, and to a lesser extent observably, found working in a chaos of tools and parts, unable to give a coherent account of what they know or how they have proceeded. In fact, some experienced technical specialists report incoherence as the principal characteristic of a technician in trouble. It seems appropriate that a failure of control should be displayed as a failure to keep order and a failure to be able to order one's words to say what is wrong, and the technicians see the connection clearly.

The counter to this threat of chaos is thought to be a systematic approach to the work. Several of the technicians involved in this study strongly suggested observing a senior technician from another team. This technician is famous for control of territory and of each task, carrying only those tools needed for a specific job, keeping all put away except those actually in use, and performing even the messiest operations with minimum fuss and in the cleanest possible fashion. During the period of my observation, certainly, both machines and customers seemed to be well under control. This, perhaps, is system in the extreme, but systematic behavior is highly valued in general. A systematic approach to troubleshooting is thought to be the best assurance that everything will get done that should be done, and that everything will be learned about the situation that can be learned. Being systematic has the advantage of being interruptible, in that one's place is known and could be resumed on a later visit. Furthermore, given the premises of the directive documentation, a systematic approach to the problem which at least covers the same material as the documentation means that failure to solve the problem is, in some sense, not the technician's alone but may be shared with the documentation. The converse is true as well: being known for an unsystematic approach to service problems may mean that success is not perceived as due to the technician's efforts but may instead be attributed to luck.

Systematic behavior includes far more than using the directive documentation. There is an unwritten body of things to be done and how to do them that "everyone knows." This includes basic things like the clearance to which one adjusts clutches, which is a specification left over from an earlier machine, and caveats that certain cables need to be separated to avoid electrical noise problems. Systematic behavior also includes somewhat less than the documentation: another part of common knowledge is all the things required by the documentation that do not really need to be done, or need to be done only in some circumstances. In these terms, the documentation is seen as a system to use if you do not know what to do on your own; it is better than no system at all.

Systematic behavior, then, is the basic requirement for a good reputation. Technical prowess is also much admired. I have continually emphasized the importance of the social dimensions of service work, and they are indeed vital. However, there is also a critical component of the work in which the technician actually has to do something to the machine, and this requires both skill and understanding. It is these purely technical procedures that in some sense define the category of technician and provide the basis for all the work done in service. The ability to do this work gains the technicians access to the customers' environment, even if it is not always what is necessary to solve the customers' problems with their machines.

The admired forms of technical expertise may consist of skill at performing difficult procedures quickly and neatly or may be seen in the development of new fixes or ways to work around the cumbersome prescribed procedures. Until they become part of what "everyone knows," such fixes may be owned by the technicians who developed them and will be associated with their names. This is not an unmixed blessing for the owner's reputation, because not all of their colleagues will have the same problems or the same need for the fixes. One technician had developed a fix for a specific bad habit of some users; technicians whose customers did not share that habit found the fix to be rather a nuisance. Other technicians may view someone's favorite fix as a way to put off the real job that needs to be done, perhaps until some other technician gets the call.

In fact, good reputations are discussed almost entirely in terms of technical skills, while bad ones seem primarily to be associated with unsystematic approaches to problems, a form of losing control. Although the need for social skills is explicitly recognized, no technician's reputation seems to be enhanced thereby.[1] The only praise heard for anything other than technical skill was for an exceptional ability to ferret out information about the machine, using connections throughout the corporation to get explanations for things that the documentation did not cover. There is good reason for this if one considers the technicians to be an occupational community as described by van Maanen and Barley (1984). There is no significant career path for technicians; the primary status within the community is that of member. One participates in the community by becoming and remaining a competent practitioner. Accordingly, such differentials as exist within the community are those of performance, or reputation for performance, and the valued performances are those that relate to the definition of a competent performer. The definition of technician centers on the skills of dealing with machines more than the actual work does, and the definition of competence or even superiority follows suit.

Social competition is also conducted in terms of technical skill. This is seen in claims to have originated certain fixes, or in contrapuntal recitations of procedures with each participant trying to show more knowledge of the difficult points, ways around them, and likely ways to fail. The finale to one such recitation was a claim to have actually done just that procedure on a specific machine, a claim not only to know the fix but to be able to do it as well. Good verbal performance seems important to the technicians, although it is unclear how much it contributes directly to their reputations. Verbal performance is the medium of their social competition; it is also the medium through which they preserve their hard-won knowledge in war stories. Moreover, it amuses them; most of the technicians I observed worked very

[1] Perhaps the technicians see these skills as part of the normal human repertoire and so not noteworthy. In fact, the skills are distributed as unevenly among technicians as among most people. One suspects it is the definition of the work, which does not, after all, originate with those who do it, that keeps these important skills from being celebrated in the same way as those involving the manipulation of machines.

hard at being funny, often at the expense of their fellows. While not as formalized as "sounding" or "the dozens" (Labov 1969; Abrahams 1974), there is an obvious pattern of challenge and response in the joking. It is never done in the presence of customers, where technician behavior is carefully controlled and most of the jokes they tell are at their own expense. The jokes they tell when offstage are often much more aggressive and are usually at each other's expense. The same technicians who support each other with advice, parts, and hours of assistance also feel free to make any sort of joke about the other members of their group.

The combination of individual, challenging work with a supportive community may be the key to the attraction this job has for the technicians. They participate as individuals, and they work independently. To a great extent they manage their own time and their own accounts. The work is sporadic and unpredictable and therefore cannot be scheduled. Each service call is potentially something new, and initially they deal with it alone. However, they have the resources of the group for support and potentially the resources of the entire corporation, if needed. In this arena they can make or lose their reputation, but no single service call will be decisive. Technicians are quite explicit about how much they value their independence. They also talk about how much they like the fact that they have to think about their work. With advantages like these, unhappy customers and erratic machines are not major drawbacks; they are opportunities to be heroic and material for better stories.

PRACTICE AND QUESTIONS OF WORK

In the introduction, I invoked the work of Williams (1983) and Wadel (1979) to indicate that in common usage the word "work" now refers to the relationship of employment much more than to either the doing or what is done, and that employment further skews our understanding of work by making what is said to be done part of a contest about reward and status. I further suggested that this emphasis may omit vital elements of working practice. I asked what might be learned by shifting the focus and concentrating on work practice instead of on relations of employment. In this study, the work practice of field service technicians is the focus, and the social relations I have discussed are those that come into being around the work. The work of field service is seen to be an important component of the worker's identity, and this importance is revealed in discourse among the workers. This discourse also reveals the work as problematic, and discussions of skill, technique, and simple coping are common within the community of technicians. The work patterns the social geography of the workers through the distribution of work sites and the allocation of responsibility for particular sites; it also divides the population of the world into those inside the community and those without, and the difficult relations across this divide are a common topic among the workers.

Other questions remaining from the introduction are whether there is a possible conflict between work as doing, as practice, and work as activities explicitly prescribed in the relationship of employment, and what might be revealed about such a conflict by a study of work practice. The work done by the technicians I studied is often very different from the methods specified by their management in the machine documentation. There is clearly a disparity between the tasks that they are told to accomplish and the means that are said to be adequate to the task. The technicians choose to give accomplishing the task priority over use of the prescribed means, and so they resolve problems in the field any way they can, apparently believing that management really wants accomplishment more than strict observation of the prescriptions for work. The technicians pay more attention to other messages from management which address the goals of service, giving the technicians a general mandate to solve problems. Managers do say, for example, that customer satisfaction is the primary goal of the corporation, and such messages can be interpreted to warrant a wide range of activities. However, the need to choose from conflicting definitions of the work and the means thereto also opens the way for continuing disputes about the very nature of the job, the legitimacy of different activities, or the adequacy of one's compensation, since it is not clear which activities require compensation as being part of one's work.

In order to understand the usefulness of studying work practice, it may be illustrative to use what has been revealed here about a particular work practice to consider some of the issues raised in the literature on work. Most of this literature seems to focus on the relationship of employment; however, this is rarely at the level of a worker, a job, and an employer. The discussion instead tends to be in terms of jobs and workers in the aggregate, for example as Labor, which has come to mean a commodity, a class, or a political movement more than work itself (Williams 1983). The common theme in the literature is a perception that the nature or circumstance of work is changing, and the concomitant question is what this will mean for workers, but the meaning looked for is a narrowly political one. De-skilling theory is one segment of this literature, headed by Harry Braverman's analysis (1974) of Taylor's Scientific Management and its development in subsequent management literature. The literature on proletarianization presumes de-skilling and seeks to analyze its effect on class structure (Crompton and Jones 1984; Marshall and Rose 1988). Another segment is the literature on the sociology of automation, which begins by noting the alienation of factory workers (Chinoy 1964; Terkel 1974, pp. 159–94) and attempts to analyze the effect of the introduction of new technologies to those factories (Blauner 1964; Gallie 1978; Zuboff 1988).

What can be said of these works, given the realities of field service that I have discussed in this study? The management of the corporation for which the technicians work has pursued a strategy of de-skilling through the use of directive documentation. This does not actually deprive the workers of the skills they have, however; it merely reduces the amount of information given to them. In fact, some of the information initially removed has been replaced in response to protests from the field, and I think the discussion of technicians' work practice makes it clear that de-skilling is not possible in any real sense. While de-skilling remains a management intention, one of the standard criticisms of de-skilling theory has been that it describes management intentions much more accurately than working realities (Kusterer 1978).

Proletarianization seems moot with reference to my study, owing to both the absence of effective de-skilling and the nebulousness of concepts of class in the United States. The sociology of automation might appear to be relevant to field service work because of its emphasis on technological developments, but it does not discuss the realities of fixing machines. Nor is this job part of the postindustrial service economy envisioned by Daniel Bell (1973). Bell's concept of the service sector is focused on health, education, research, and government; the workers therein are office workers, bureaucrats, and scientists. This formulation simply does not address the issue of where machines come from or who will fix them; both functions presumably are part of the remains of the old manufacturing economy. This is the main problem with all this literature. It is not well grounded in analysis of work practice, so its presumptions and prescriptions of what is to be done are not based on what is done and what needs to be done, on the reality of the job, the task to be accomplished.

"The specialization of work to paid employment . . . is the result of the development of capitalist productive relations" (Williams 1983, p. 335). Anthropologists studying work in regions where capitalism is not the only or perhaps even the dominant element in the economy have described the ways in which certain kinds of work become a principal way of defining one's place in society; Enid Schildkrout, David Parkin, Anthony Cohen, and Mary Searle-Chatterjee, all in Sandra Wallman's 1979 collection *The Social Anthropology of Work*, provide examples of work contributing to both individual and collective identities. A concept of collective identity that is more closely linked to modern industrial work is van Maanen and Barley's definition of occupational community: "Occupational communities represent bounded work cultures populated by people who share similar identities and values that transcend specific organizational settings. Moreover, self-control is a prominent cultural theme in all occupational communities, although its realization is highly problematic" (van Maanen and Barley 1984, pp. 314–15). These authors see control of an occupational community as a clear challenge to management. The fragmentation of work caused by the more successful attempts at de-skilling has increased management's control at the expense of the community. A different approach for management is to offer career development which will increase ties to the organization rather than to the occupational community. This strategy depends on the possibility of real career movement, and the community may counter it by devaluing promotions. Occupational communities can be expected to resist changes in the work process, particularly ones that increase their ties to the organization or that appear to be aimed at de-skilling the practitioners.

Van Maanen and Barley's concept of the occupational community and the examples mentioned from the literature on work in other societies all show the remarkable range of ways in which work can contribute to the definition of the worker's identity, either singly or as a community. What is almost completely missing from most of the studies discussed so far is a focus on the work itself. What role do the events of a day's work play in the process of defining identity? What are the relationships between the work and the worker, between the workers in the presence of the work, or between the workers and the consumers of the work? Most anthropological investigations of work that consider practice have done so in nonindustrial contexts; when one turns to the literature on industrial wage labor, interest is found in issues around work but not in the practice itself. Authors may presume that modern jobs are in some sense known, perhaps from the formal job descriptions that Wadel called into question. This black box treatment of modern occupations denies that there is anything interesting or problematic about the work itself; surely one goal of industrial anthropology must be to open these black boxes and see if the goings-on within are those we expect.

Obviously, I consider that wage work is still problematic in a mature capitalist society, and that daily working practice in industrial jobs is a rewarding subject for ethnographies of work. In this study of work practice I have attempted to focus more on how the work is done than on how it is supposed to be done. This approach does not escape the relationship of employment, since employers do affect the way their employees work, but it allows one to focus on those practices employed in the performance of the task without worrying so much about whether the official definition of the work includes them. This is not unique, and it is worth comparing the work of field service with other ethnographies of work.

Ken Kusterer (1978) investigated work practices and work relations to identify the skills required in relatively unskilled jobs such as bank teller or attendant to a machine making glued paper cones used in fast food restaurants. The most interesting point of comparison between these workers and the technicians we have been considering is Kusterer's division of the workers' knowledge into three broad categories: knowledge of routine processes, knowledge of potential variations, and social knowledge. The routine processes comprise what is defined as "the job" by management and will probably be the main subject of any training. Some of the routine procedures he found had been improvised by the workers, but these were only sometimes recognized by management as part of the work. Potential variations include those in the machines used in the work and those in the material of the work; social knowledge includes customers, coworkers, and managers.

This working knowledge divides further into three more categories of varying interest to the workers. The basic set comprises those things that happen routinely on the job and that "everyone knows." The set that is interesting to the workers includes those things that happen with some regularity and with some effect on the work. These can be learned, and it is perceived to be advantageous to do so. Things that occur still more rarely appear unique, accidental, and unknowable; Kusterer says these hold no interest for his informants. Perhaps their rarity makes it impossible to learn from them. He notices that workers with bad reputations tend to regard as unique and accidental those events that their colleagues think are knowable. Conversely, the value of experienced workers is their ability sometimes to see as regular and learnable those events otherwise viewed as unique and accidental. This patterning of working knowledge matches that of field service technicians, except the technicians cannot ignore the very rare events; those problems must be solved too. In fact, the rare events fascinate them, since new problems combine the possibility of failure, the opportunity for heroism, and the probability of a good story.

Kusterer observes that the process of working and learning together creates a situation that the workers value, and they resist having it disrupted by their employers through events such as a reorganization of the work. This resistance can surprise employers who think of labor as a commodity that can be arranged to suit their ends. The problem for the workers is that this community which they have created was not part of the series of discrete employment agreements by which the employer populated the workplace, nor is the role of the community in doing the work acknowledged. The work can only continue free of disruption if the employer can be persuaded to see the community as necessary to accomplishing the work.

If the workers in Kusterer's study display more solidarity than might be imagined by their management, other studies reveal rather more diversity than expected. Sailors on Great Lakes' freighters are thought to be the same as sailors on ocean-going vessels by management, union organizers, and merchant marine academies alike; the sailors beg to differ. M. Estellie Smith (1977) reports that sailors of either variety can explain in detail why they are real sailors and the others are not. The significant point is that discovering and accepting this difference clarifies other differences that seem inexplicable under the assumption that sailors are all the same.

Some issues are known to be important to workers in certain jobs, such as time for railroaders, but this knowledge has a deceptive simplicity. Time turns out to be a complex concept with a wide range of variation when examined in practice. L. S. Kemnitzer (1977) describes some of the multiple ways in which time is used or perceived by railroaders, few of which correspond exactly to time as registered by clock or calendar. A railroader's career is an accumulation of time on the job, not just the passage of years. Keeping schedule for a given train is not simply a matter of clocks and therefore distance but includes maintaining the required intervals between trains while considering their speed differentials and differing priorities, keeping track of the time remaining in the shift, and knowing the distances and times required to accommodate these different constraints by getting to the next stop or the next siding to get off work within the allotted time or to clear the track for a train with priority. Time is also rhythm, essential to switching practice, and as such is a skill, part of the repertoire of a good railroader, and it governs the railroaders' actions in ways that cannot be prescribed by the company. It is also a weapon in struggles with those companies: railroaders are not allowed to strike, but working to rule, depriving the company of the benefit of their skills at timing operations, essentially makes the railroad inoperable.

The significant thing about these studies is that the examination of practice reveals a complexity that cannot be seen from a distance; this complexity constrains how the work can be done and therefore has crucial implications for those making policy about work. Sailing a ship is not the same on the ocean as on the Great Lakes; time for railroaders is not simply a matter of having an accurate timepiece. Similarly, field service work is not just a matter of fixing broken machines but requires the maintenance of a triangular relationship between customers, machines, and technicians.

Police work practice has been described in more detail than most jobs in the United States, perhaps due to the frequency of questions about the role of police in our society. Jonathan Rubinstein (1973) and van Maanen (1973, 1979, 1980, 1984, 1988) discuss in detail the process of becoming a police officer and of doing police work, and they examine the daily interactions between those in the community of police officers and those outside, simultaneously their customers, their employers, and the subjects of their work. Rubinstein explains why understanding police work requires a detailed description of practice, an explanation applicable to other kinds of work as well:

> Like every other kind of work, police work generates demands on the people who do it and encourages them to develop skills and techniques for making the job easier. It may be good or bad work, but it is work, and before any judgments of its moral character or suggestions for reforming it can be made, the work itself must be described. This has not been done. The reporter is always an outsider whose access to the police is assured by his pledge not to reveal what he knows of police work. Scholars rarely have either the time or the inclination to seek close ties with the men they want to study. Instead of studying the work, they report on its organization and administration; instead of describing what the men do, they

examine their feelings and values. These may be worthwhile things to do, but they cannot be done properly unless the observer understands the nature of the work whose administration he is examining, and the constraints and contingencies which affect the men who do it. (Rubinstein 1973, pp. x–xi)

There are numerous aspects of police work that show some curious parallels to the work of field service, and there are some crucial differences. The most obvious difference has to do with the potential for violence, which seems to be inevitable in police work and does not exist in service work, and it seems probable that this difference will give even remarkably similar behaviors a different character in the two worlds. Another striking difference is found around the question of sharing work information. Both ethnographers report that policemen cheerfully share information on street practice, particularly when breaking in newly trained officers. Rubinstein, however, reports that apart from this, the police he observed were very reluctant to share information about what was happening on the street. Such information can be used to the officer's personal advantage, while its disclosure will at best advance someone else's career, and at worst lead to dismissal. Technicians, however, share what they know.

The parallels begin in training: both police and technicians start their respective jobs with a period of mandatory training that is generally discounted on the street or in the field. Both groups insist that the job is really learned through practice, and in both jobs the war stories of actual experiences in the field are an important aspect of communicating norms of practice to new members. One aspect of practice vital to each is learning what to notice on the job, to see what is important and significant for the problem at hand. Similarly, both have to listen to people outside the work group and must learn how to interpret what they say. Outsiders are not to be believed implicitly, but their words must be given some credence while ultimate judgment is reserved. Service technicians have the advantage here, in that they work with fewer people, and credibility or even a common language can be negotiated over time.

Another parallel between the work of the police and that of

service technicians is the unpredictability of the working day, in which almost anything might happen, while there is a considerably smaller set of events that may reasonably be expected, and normal days can be somewhat boring. The extremes, of course, are rather worse for police. Practitioners in both domains are relatively autonomous, working alone or in pairs and away from immediate supervision; they are responsible for having what they need for the job with them. Both groups are difficult for their managers to control, owing to the nature of the work itself, and in each case their immediate managers have found a compromise between the need to exercise some control, or to be seen exercising control, and the need to encourage the autonomy and self-reliance that will actually get the job done when the police officer or field service technician is out in the field alone. My observations of this among the technicians match the reports from the police. From Rubinstein's analysis and my observations, it also appears to be true for both groups that higher levels of management intend that those in the field be more closely controlled and often introduce programs to implement those intentions. The tasks to be done in both police work and service work are, however, intractable, not amenable to control by management programs, and the challenge for the lowest level of management is to balance the field realities against the expectations of higher management.

Ultimately the two jobs diverge. Police work, in the end, is about people, watching them, seeing them, controlling them, catching them. Service work is about machines and people, and some of the vital skills involve doing things to machinery. At that point, work is mechanical practice: the hand's skills and the mind's understanding of and sympathy for a machine.

Such practice is the subject of Douglas Harper's study (1987) of an upstate New York shop run by Willie, a man who is an automobile mechanic, a farm machinery mechanic, a blacksmith. Willie can repair or fabricate virtually anything made of metal, drawing on a vast collection of old cars and machinery for parts or raw materials. He is a literal manifestation of Lévi-Strauss's metaphoric *bricoleur* (1966). Harper grounds this ethnography in detailed descriptions of several projects Willie does, focusing on the array of skills employed and the variety of knowledge necessary to choose the most appropriate method for the task. He discusses the importance of kinesthetics—knowing how various mechanisms and certain actions feel—and he emphasizes the importance of an overall understanding of the machine if one is to understand the interactions of the mechanism and possibly find the deeper roots of problems. There is an attention to process; Willie is concerned with doing things correctly, not merely getting them done.

Harper describes the community that centers on Willie's shop, which variously resembles a market, a school, and a cooperative. However, the community does not bring people to the shop, not even Harper, a sociologist; Willie's skills are the reason for this community. Ultimately, someone has to fix the machines, and Willie has the skill to do so. The essence of Harper's ethnography, which is the point of the other works cited here and the intent of this work, is to describe skilled practice, to show how varied and demanding is the work when seen in detail, and then to show how other things develop from that practice. In any sort of work, something has to get done, and it is that accomplishment which makes possible all else that work accomplishes in society. Discussions of work that omit this vital aspect of practice lose the point from which anything else that may be described originates.

CODA

There is an existential dilemma at the heart of service: the technicians are responsible in a world in which they have very little control. Their job is to respond to trouble and to anticipate and avoid trouble when they can, but the setting in which they perform is largely constructed by other people, is inhabited by other people, and is inherently unpredictable. The design and manufacture of the machines is done with goals that may include self-diagnostic capabilities, few and simple adjustments, and easy repair or replacement of all subassemblies, but these are often displaced by other goals: reduction of the unit cost or adaptation to a particular method of manufacturing. The design and writing of the documentation and the design and implementation of the training are done outside the community of technicians by persons whose perspective may be shaped less by the practice of service than by constraints of the documentation or training communities, or by corporate policy defining the tasks of service work. The matching of machine capabilities to customers' needs and intended uses is done by salespeople whose interest in making the sale may override their concern for whether the machine will be adequate to the customers' needs. The actual place in which the work is done is the customer's business site, in which the customers use the machines in their own ways and to their own ends. The other players in this arena are equally beyond the technicians' control: customers, fellow technicians, and management on both sides come and go, pursuing their own agendas while participating in the constitution of the service situation.

Because one cannot predict when or how a given machine will fail, technicians cannot plan their work in any detail. If they have time, they can try to anticipate failure and schedule maintenance, but normally they can only react to a situation already defined as a problem. When they are on a service call, it is not always apparent whether the machine in question has actually failed, or how it has failed. The immediate stimulus for the service call is a perception of a machine problem by some members of the user community, but this same user community may misconstrue the state of the machine, from the technicians' point of view, or be unable to represent it in terms meaningful to the technicians. The result is that both the definition and resolution of the problem may require negotiation between the users and the technician.

This describes the uncertainties at the level of the service call. A diachronic perspective suggests other complicating factors. The machine as a type has a life span, with different problems known to occur at different points along the span and various modifications mandated or projected in response to known problems. The machine as an individual has a history of known problems, known repairs, and modifications done or not. The history includes which technician had the machine as an assigned responsibility, which technicians did which repairs or which modifications, and whether there were periods in which the machine was not assigned to a specific technician. The social situation has its own character and history, including the users' willingness to learn the peculiarities of the machine, the types of work to be done with the machine, and the users' tolerance for disruption. Finally, the nature and continuity of the relationship between the technician and the customers, particularly the customer primarily responsible for the machine, has a critical effect on its maintenance and repair.

In this unpredictable, uncontrollable world, understanding is as problematic as control, and the tenuous nature of understanding is a threat to control. The documentation is not necessarily accurate in what it says and may say nothing at all about the problem at hand. The machine's error codes may not mean what they say. The customer may be unable to describe the problem in any significant way. Even if all of these information resources work as intended, the facts may not constitute a coherent picture of the machine. To preserve the understandings they create from such situations, the technicians pool their knowledge. They share what they know, telling each other about new fixes they have found or strange new problems they have encountered. When they have been working on each other's machines, they tell what they have done and what the machine needs. Given that they all work on each other's machines, there is no incentive to keep information private and every reason to see that the other technicians have all relevant knowledge.

There are positive aspects to the unpredictability of the work. It gives the technicians a certain autonomy in scheduling, since the time required to fix the machine cannot be known until the machine's problems are determined. Because the diagnostics or documentation do not and cannot anticipate all the possible failures, the technicians will have to think for themselves in solving some problems. The real benefit of this situation for the technicians is that they are needed. Their errand is one of rescue, and this uncertain, perilous world affords the opportunity for heroism of a sort. In this context, the "technician" is defined as someone who fixes the world and makes it right, and the technicians cultivate the image of white-hatted wrench-slingers. They value their job both for the challenge of the work and for the identity as hero. Competent practice at the job creates the identity, and their stories celebrate both the practice and the heroism while preserving the details of the practice and so helping to perpetuate the identity.

This study sees the skilled practice of field service work as necessarily improvised, at least in diagnosis, and centered on the creation and maintenance of control and understanding. Control and understanding are achieved through a coherent account of the situation, requiring both diagnostic and narrative skills. Understanding is maintained through circulation of this knowledge by retelling the narratives to other members of the community, and this preservation of understanding contributes to the maintenance of control. Practice also depends on the technical skills necessary to do any repair, the only element of technicians' practice commonly acknowledged to be part of the work. Technicians, however, know that control requires more than purely mechanical skills.

Accordingly, when technicians gather, their conversation is full of talk about machines. This talk shows their understanding of the world of service; in another sense, the talk creates that world and even creates the identities of the technicians themselves. But neither talk nor identity is the goal of technicians' practice. The goal is getting the job done, keeping the customers happy, and keeping the machines running. Talking about machines, a vital element of their practice, is for the technicians simply a means to this end.

References

Abrahams, Roger D. 1974. Black Talking on the Streets. In *Explorations in the Ethnography of Speaking*, ed. Richard Bauman and Joel Sherzer. Cambridge: Cambridge University Press.

Akrich, Madelaine. 1992. The De-Scription of Technical Objects. In *Shaping Technology/Building Society*, ed. W. E. Bijker and J. Law. Cambridge: MIT Press.

Bell, Daniel. 1973. *The Coming of Post-Industrial Society*. New York: Basic Books.

Bennett, W. Lance, and Martha S. Feldman. 1981. *Reconstructing Reality in the Courtroom*. New Brunswick, N.J.: Rutgers University Press.

Blauner, Robert. 1964. *Alienation and Freedom*. Chicago: University of Chicago Press.

Blomberg, Jeanette L. 1987. Social Interaction and Office Communication: Effects on User Evaluation of New Technologies. In *Technology and the Transformation of White-Collar Work*, ed. Robert Kraut. Hillsdale, N.J.: Erlbaum Press.

Braverman, Harry. 1974. *Labor and Monopoly Capital: The Degradation of Work in the Twentieth Century*. New York: Monthly Review Press.

Chinoy, Ely. 1964. Manning the Machines—The Assembly-Line Worker. In *The Human Shape of Work: Studies in the Sociology of Occupations*, ed. Peter Berger. New York: Macmillan.

Cohen, Anthony P. 1979. The Whalsay Croft: Traditional Work and Customary Identity in Modern Times. In *Social Anthropology of Work*, ed. Sandra Wallman. ASA Monograph 19. London: Academic Press.

Crompton, Rosemary, and Gareth Jones. 1984. *White-Collar Proletariat: Deskilling and Gender in Clerical Work*. London: Macmillan.

Early, Evelyn. 1982. The Logic of Well Being: Therapeutic Narratives in Cairo, Egypt. *Science and Medicine* 16: 1491–97.

Evans-Pritchard, E. E. 1951. *Kinship and Marriage among the Nuer*. Oxford: Oxford University Press.

Gallie, Duncan. 1978. *In Search of the New Working Class: Automation and Social Integration within the Capitalist Enterprise*. Cambridge: Cambridge University Press.

Garfinkel, Harold. 1967. *Studies in Ethnomethodology*. Englewood Cliffs, N.J.: Prentice-Hall.

Geertz, Clifford. 1973. *The Interpretation of Cultures*. New York: Basic Books.

——. 1983. *Local Knowledge*. New York: Basic Books.

Harper, Douglas. 1987. *Working Knowledge: Skill and Community in a Small Shop*. Chicago: University of Chicago Press.

Jordan, Brigitte. 1987. *Modes of Teaching and Learning: Questions Raised by the Training of Traditional Birth Attendants*. IRL Report No. IRL87-0004. Palo Alto: Institute for Research on Learning. Earlier version delivered at the annual meeting of the American Anthropological Association, Philadelphia, 1986.

Kemnitzer, L. S. 1977. Another View of Time and the Railroader. *Anthropological Quarterly* 50(1): 25–29.

Kusterer, Ken C. 1978. *Know-How on the Job: The Important Working Knowledge of "Unskilled" Workers*. Boulder, Colo.: Westview Press.

Labov, William. 1969. Rules for Ritual Insults. In *Studies in Social Interaction*, ed. D. Sudnow. New York: Macmillan.

Latour, Bruno. 1986. How to Write "The Prince" for Machines As Well As for Machinations. A working paper for the seminar on Technology and Social Change, Edinburgh, June 12–13.

——. 1988. Sociology of a Door. *Social Problems* 35(3): 298–310. Pseudonymously attributed to Jim Johnsons, Columbus Ohio School of Mines.

Lave, Jean. 1988. *Cognition and Practice: Mind, Mathematics, and Culture in Everyday Life*. Cambridge: Cambridge University Press.

Lévi-Strauss, Claude. 1966. *The Savage Mind*. Chicago: University of Chicago Press.

Marshall, G., and D. Rose. 1988. Proletarianisation and the British Class Structure. *British Journal of Sociology* 39(4): 498–518.

Martin, Joanne. 1982. Stories and Scripts in Organizational Settings. In *Cognitive Social Psychology*, ed. Albert H. Hastorf and Alice M. Isen. New York: Elsevier/North-Holland.

Myerhoff, Barbara. 1986. "Life Not Death in Venice": Its Second Life. In *The Anthropology of Experience*, ed. Victor W. Turner and Edward M. Bruner. Urbana: University of Illinois Press.

News & Digest. 1988. 13(3). Rochester: Xerox.

Parkin, David. 1979. The Categorization of Work: Cases from Coastal Kenya. In *Social Anthropology of Work*, ed. Sandra Wallman. ASA Monograph 19. London: Academic Press.

Radcliffe-Brown, A. R. 1950. Introduction. In *African Systems of Kinship and Marriage*, ed. A. R. Radcliffe-Brown and Daryll Forde. Oxford: Oxford University Press.

Rosaldo, Renato. 1986. Ilongot Hunting as Story and Experience. In *The Anthropology of Experience*, ed. Victor W. Turner and Edward M. Bruner. Urbana: University of Illinois Press.

Rubinstein, Jonathan. 1973. *City Police*. New York: Farrar, Straus, and Giroux.

Sacks, Harvey. 1970. Unpublished Lecture Notes, Spring.

——. 1972. On the Analyzability of Stories by Children. In *Directions in Sociolinguistics*, ed. John Gumperz and Dell Hymes. New York: Holt, Rinehart and Winston.

——. 1974. An Analysis of the Course of a Joke's Telling in Conversation. In *Explorations in the Ethnography of Speaking*, ed. Richard Bauman and Joel Sherzer. Cambridge: Cambridge University Press.

Schildkrout, Enid. 1979. Women's Work and Children's Work: Variations among Moslems in Kano. In *Social Anthropology of Work*, ed. Sandra Wallman. ASA Monograph 19. London: Academic Press.

Schon, Donald A. 1983. *The Reflective Practitioner: How Professionals Think in Action*. New York: Basic Books.

Searle-Chatterjee, Mary. 1979. The Polluted Identity of Work: A Study of Benares Sweepers. In *Social Anthropology of Work*, ed. Sandra Wallman. ASA Monograph 19. London: Academic Press.

Smith, Barbara Herrnstein. 1980. Narrative Versions, Narrative Theories. *Critical Inquiry* 7(1): 213–36.

Smith, M. Estellie. 1977. Don't Call My Boat a Ship! *Anthropological Quarterly* 50(1): 9–17.

Suchman, Lucy. 1987. *Plans and Situated Actions: The Problem of Human-Machine Communication*. Cambridge: Cambridge University Press.

Terkel, Studs. 1974. *Working*. New York: Pantheon Books.

van Maanen, John. 1973. Observations on the Making of Policemen. *Human Organization* 32: 407–18.

——. 1979. The Fact of Fiction in Organizational Ethnography. *Administrative Science Quarterly* 24: 539–50.

——. 1980. Beyond Account. *Annals of American Political and Social Science* 58: 458–63.

——. 1984. Making Rank. *Urban Life* 13: 155–76.

——. 1988. *Tales of the Field: On Writing Ethnography*. Chicago: University of Chicago Press.

van Maanen, John, and Stephen R. Barley. 1984. Occupational Communities: Culture and Control in Organizations. In *Research in Organizational Behavior*, Vol. 6., ed. Barry M. Staw and L. L. Cummings. Greenwich, Conn.: JAI Press.

Wadel, Cato. 1979. The Hidden Work of Everyday Life. In *Social Anthropology of Work*, ed. Sandra Wallman. ASA Monograph 19. London: Academic Press.

Wallman, Sandra, ed. *Social Anthropology of Work.* Association of Social Anthropologists, Monograph 19. London: Academic Press.

Williams, Raymond. 1983. *Keywords: A Vocabulary of Culture and Society.* London: Fontana Press.

Zuboff, Shoshana. 1988. *In the Age of the Smart Machine: The Future of Work and Power.* New York: Basic Books.

Index